Functional Safety

A Straightforward Guide to applying IEC 61508 and Related Standards

Second edition

David J Smith
BSc, PhD, CEng, FIEE, FIQA, HonFSaRS, MIGasE

Kenneth G L Simpson
MPhil, FIEE, FInstMC, MIGasE

ELSEVIER
BUTTERWORTH HEINEMANN

AMSTERDAM • BOSTON • HEIDELBERG • LONDON • NEW YORK • OXFORD
PARIS • SAN DIEGO • SAN FRANCISCO • SINGAPORE • SYDNEY • TOKYO

Elsevier Butterworth-Heinemann
Linacre House, Jordan Hill, Oxford OX2 8DP
200 Wheeler Road, Burlington, MA 01803

First published 2001
Second edition 2004

British Library Cataloguing in Publication Data
A catalogue record for this book is available from the British Library

Library of Congress Cataloguing in Publication Data
A catalogue record for this book is available from the Library of Congress

ISBN 0 7506 6269 7

For information on all Elsevier Butterworth-Heinemann publications visit our website at http://books.elsevier.com

Printed and bound in Great Britain

Contents

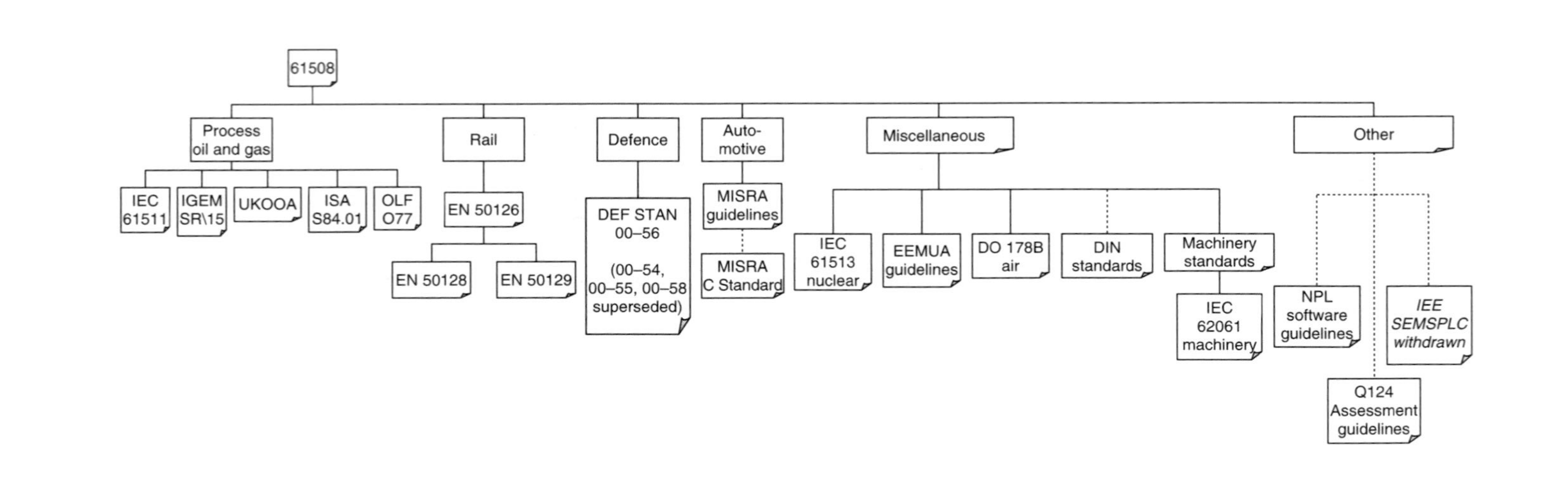

61508
Process oil and gas
IEC 61511
IGEM SR\15
UKOOA
ISA S84.01
OLF O77
Rail
EN 50126
EN 50128
EN 50129
Defence
DEF STAN 00–56
(00–54, 00–55, 00–58 superseded)
Auto-motive
MISRA guidelines
MISRA C Standard
Miscellaneous
IEC 61513 nuclear
EEMUA guidelines
DO 178B air
DIN standards
Machinery standards
IEC 62061 machinery
Other
NPL software guidelines
Q124 Assessment guidelines
IEE SEMSPLC withdrawn

A Quick Overview

Functional safety involves identifying specific hazardous failures which lead to serious consequences (e.g. death) and then establishing maximum tolerable frequency targets for each mode of failure. Equipment whose failure contributes to each of these hazards is identified and usually referred to as 'safety-related'. Examples are industrial process control systems, process shutdown systems, rail signalling equipment, auto-motive controls, medical treatment equipment etc. In other words, any equipment (with or without software) whose failure can contribute to a hazard is likely to be safety-related.

Since the publication of the first edition of this book, in 2001, the application of IEC 61508 has spread rapidly through most sectors of industry. Also, the process sector IEC 61511 has been published. The opportunity has therefore been taken to update and enhance this book in the light of the authors' recent experience. Chapter 5 is now devoted to IEC 61511 and Chapters 13 and 14 have been added to provide even more examples.

The maximum tolerable failure rate for each hazard will lead us to an integrity target for each piece of equipment, depending upon its relative contribution to the hazard in question. These integrity targets are known as 'safety-integrity levels' and are usually described by one of four discrete bands described in Chapter 1.

SIL 4: the highest target and most onerous to achieve, requiring state of the art techniques (usually avoided)
SIL 3: less onerous than SIL 4 but still requiring the use of sophisticated design techniques

SIL 2:	requiring good design and operating practice to a level not unlike ISO 9000
SIL 1:	the minimum level but still implying good design practice
<SIL 1:	referred to (in IEC 61508 and other documents) as 'not-safety related' in terms of compliance

An assessment of the design, the designer's organisation and management, the operator's and the maintainer's competence and training should then be carried out in order to determine if the proposed (or existing) equipment actually meets the target SIL in question. The steps involve:

Setting the SIL targets	Chapter 2.2
Capability to design for functional safety	Chapter 2.1
Quantitative assessment	Chapters 3, 5, 6 & 7
Qualitative assessment	Chapters 3, 4 & 5
Establishing competency	Chapter 2.1
As low as reasonably practicable	Chapter 2.3
Reviewing the assessment itself	Appendix 2

IEC 61508 is a generic standard which deals with the above. It can be used on its own or as a basis for developing industry sector specific standards (Chapter 9). In attempting to fill the roles of being both a global template for the development of application specific standards, and being a standard in its own right, it necessarily leaves much to the discretion and interpretation of the user. Plans to revise it are well under way and a draft is planned for June 2004 with a target of 2006 for finalisation. It is now a BS EN document.

It is vital to bear in mind, however, that no amount of assessment will lead to enhanced integrity unless the assessment process is used as a tool during the design-cycle.

NOW READ ON!

Acknowledgements

The authors are very grateful to Mike Dodson, Independent Consultant, of Solihull, for extensive comments and suggestions and for a thorough reading of the manuscript:

We are also grateful to Colin Sellers of AEA Technology Rail for inputs concerning rail related standards and UKOOA (United Kingdom Offshore Operators Association) for permission to reproduce the risk graph.

Thanks are also due to Graham Ottley of Silveretch International for many comments.

Thanks also to Mr Roger Stillman of SIRA Certification Services and to Dr Brian Wichmann for comments on the original proposals and to Dr Tony Foord for assistance with Chapter 14.

Part A

The Concept of Safety-Integrity

In this first chapter we will introduce the concept of functional safety, expressed in terms of safety integrity levels. It will be placed in context, along with risk assessment, likelihood of fatality and the cost of conformance.

The life-cycle approach, together with the basic outline of IEC 61508, will be explained.

CHAPTER 1

THE MEANING AND CONTEXT OF SAFETY-INTEGRITY TARGETS

1.1 Risk and the need for safety targets

There is no such thing as zero risk. This is because no physical item has a zero failure rate, no human being makes zero errors and no piece of software design can foresee every possibility.

Nevertheless public perception of risk, particularly in the aftermath of a major incident, often calls for the zero risk ideal. However, in general most people understand that this is not practicable as can be seen from the following examples of everyday risk of death from various causes:

All causes (mid-life including medical)	1×10^{-3} pa
All accidents (per individual)	5×10^{-4} pa
Accident in the home	4×10^{-4} pa
Road traffic accident	6×10^{-5} pa
Natural disasters (per individual)	2×10^{-6} pa

Therefore the concept of defining and accepting a tolerable risk for any particular activity prevails.

The actual degree of risk considered to be tolerable will vary according to a number of factors such as the degree of control one has over the circumstances, the voluntary or involuntary nature of the risk, the number of persons at risk in any one incident and so on. This partly explains why the home remains one of the highest areas of risk to the individual in everyday life since it is there that we have control over what we choose to do and are therefore prepared to tolerate the risks involved.

A safety technology has grown up around the need to set target risk levels and to evaluate whether proposed designs meet these targets be they process plant, transport systems, medical equipment or any other application.

In the early 1970s people in the process industries became aware that, with larger plants involving higher inventories of hazardous material, the practice of learning by mistakes (if indeed we do) was no longer acceptable. Methods were developed for identifying hazards and for quantifying the consequences of failures. They were evolved largely to assist in the decision-making process when developing or modifying plant. External pressures to identify and quantify risk were to come later.

By the mid-1970s there was already concern over the lack of formal controls for regulating those activities which could lead to incidents having a major impact on the health and safety of the general public. The Flixborough incident in June 1974, which resulted in 28 deaths, focused UK public and media attention on this area of technology. Many further events, such as that at Seveso (Italy) in 1976 through to the Piper Alpha offshore disaster and more recent Paddington (and other) rail incidents, have kept that interest alive and have given rise to the publication of guidance and also to legislation in the UK.

The techniques for quantifying the predicted frequency of failures are just the same as those previously applied to plant availability, where the cost of equipment failure was the prime concern. The tendency in the last few years has been for more rigorous application of these techniques (together with third party verification) in the field of hazard assessment. They include Fault Tree Analysis, Failure Mode and Effect Analysis, Common Cause Failure Assessment and so on. These will be addressed in Chapters 6 and 7.

Hazard assessment of process plant, and of other industrial activities, was common in the 1980s but formal guidance and standards were rare and somewhat fragmented. Only Section 6 of the Health and Safety at Work Act 1974 underpinned the need to do all that is reasonably practicable to ensure safety. However, following the Flixborough disaster, a series of moves (including the Seveso directive) led to the CIMAH (Control of Industrial Major Accident Hazards) regulations, 1984, and their revised COMAH form (Control Of Major Accident

Hazards) in 1999. The adoption of the Machinery Directive by the EU, in 1989, brought the requirement for a documented risk analysis in support of CE marking.

Nevertheless, these laws and requirements do not specify how one should go about establishing a target tolerable risk for an activity, nor do they address the methods of assessment of proposed designs nor provide requirements for specific safety-related features within design.

The need for more formal guidance has long been acknowledged. Until the mid-1980s risk assessment techniques tended to concentrate on quantifying the frequency and magnitude of consequences of given risks. These were sometimes compared with loosely defined target values but, being a controversial topic, these targets (usually in the form of fatality rates) were not readily owned up to or published.

EN 1050 (Principles of risk assessment), in 1996, covered the processes involved in risk assessment but gave little advice on risk reduction. For machinery control EN 954-1 (Safety related parts of control systems) provided some guidance on how to reduce risks associated with control systems but did not specifically include PLCs (programmable logic controllers) which were separately addressed by other IEC (International Electrotechnical Commission) and CENELEC (European Committee for Standardisation) documents.

The proliferation of software during the 1980s, particularly in real time control and safety systems, focused attention on the need to address systematic failures since they could not necessarily be quantified. In other words whilst hardware failure rates were seen as a credibly predictable measure of reliability, software failure rates were generally agreed not to be predictable. It became generally accepted that it was necessary to consider qualitative defences against systematic failures as an additional, and separate, activity to the task of predicting the probability of so-called random hardware failures.

In 1989, the HSE (Health and Safety Executive) published guidance which encouraged this dual approach of assuring functional safety of programmable equipment. This led to IEC work, during the 1990s, which culminated in the International Safety Standard IEC 61508 – the main subject of this book. The IEC Standard is concerned with electrical, electronic and

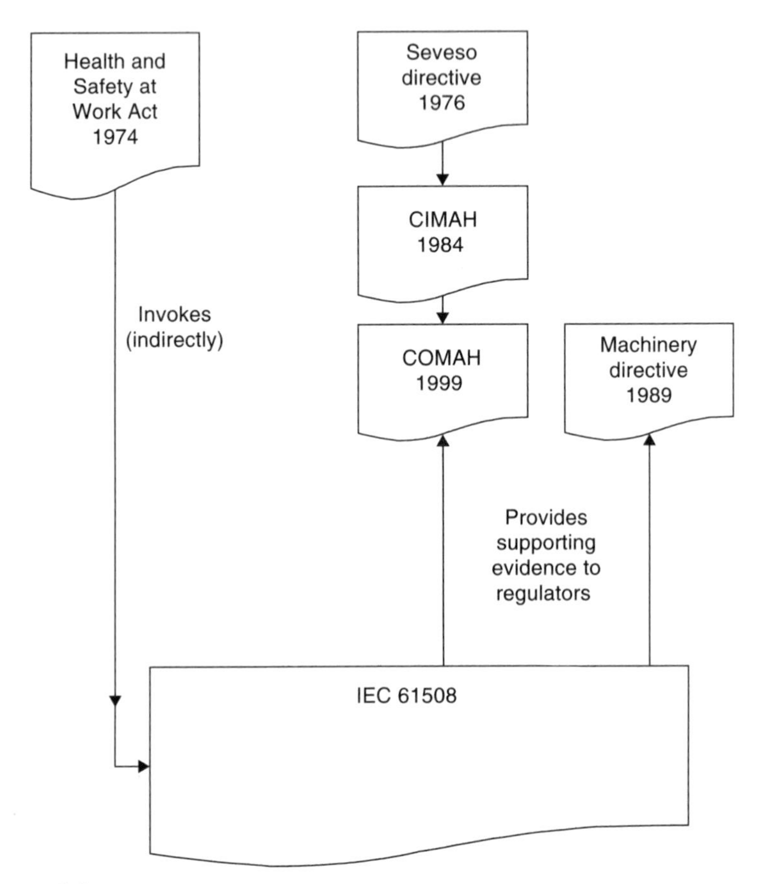

Figure 1.1

programmable safety-related systems where failure will affect people or the environment. It has a voluntary, rather than legal, status in the UK but it has to be said that to ignore it might now be seen as 'not doing all that is reasonably practicable' in the sense of the Health and Safety at Work Act and a failure to show 'due diligence'. As use of the Standard becomes more and more widespread it can be argued that it is more and more 'practicable' to use it. Figure 1.1 shows how IEC 61508 relates to some of the current legislation.

The purpose of this book is to explain, in as concise a way as possible, the requirements of IEC 61508 and the other industry-related documents (some of which are referred to as second tier guidance) which translate the requirements into specific application areas.

The Standard, as with most such documents, has considerable overlap, repetition, and some degree of ambiguity, which places the onus on the user to make interpretations of the guidance and, in the end, apply his/her own judgement.

The question frequently arises as to what is to be classified as safety-related equipment. The term 'safety-related' applies to any hardwired or programmable system where a failure, singly or in combination with other failures/errors, could lead to death, injury or environmental damage. The terms 'safety-related' and 'safety-critical' are often used and the distinction has become blurred. 'Safety-critical' has tended to be used where failure alone, of the equipment in question, leads to a fatality or increase in risk to exposed people. 'Safety-related' has a wider context in that it includes equipment in which a single failure is not necessarily critical whereas coincident failure of some other item leads to the hazardous consequences.

A piece of equipment, or software, cannot be excluded from this safety-related category merely by identifying that there are alternative means of protection. This would be to pre-judge the issue and a formal safety integrity assessment would still be required to determine whether the overall degree of protection is adequate.

1.2 Quantitative and qualitative safety targets

In the previous section we introduced the idea of needing to address safety-integrity targets in two ways:

Quantitatively: where we predict the frequency of hardware failures and compare them with some tolerable risk target. If the target is not satisfied then the design is adapted (e.g. provision of more redundancy) until the target is met.

Qualitatively: where we attempt to minimise the occurrence of systematic failures (e.g. software errors) by applying a variety of defences and design disciplines appropriate to the severity of the tolerable risk target.

The question arises as to how a safety-integrity target can be expressed in such a way as to be consistent with both approaches. During the 1990s the concept of safety-integrity levels (known as SILs) evolved and is used in the majority of documents in this area. The concept is to divide the 'spectrum' of integrity into a number of discrete levels (usually four) and then to lay down requirements for each level. Clearly, the higher the SIL then the more stringent become the requirements. In IEC 61508 (and in most other documents) the four levels are defined in Table 1.1.

Table 1.1 Safety-Integrity Levels (SILs)

Safety-Integrity Level	*High demand rate (Dangerous failures/hr)*	*Low demand rate (Probability of failure on demand)*
4	$\geqslant 10^{-9}$ to $< 10^{-8}$	$\geqslant 10^{-5}$ to $< 10^{-4}$
3	$\geqslant 10^{-8}$ to $< 10^{-7}$	$\geqslant 10^{-4}$ to $< 10^{-3}$
2	$\geqslant 10^{-7}$ to $< 10^{-6}$	$\geqslant 10^{-3}$ to $< 10^{-2}$
1	$\geqslant 10^{-6}$ to $< 10^{-5}$	$\geqslant 10^{-2}$ to $< 10^{-1}$

Note that had the high demand SIL bands been expressed as 'per annum' then the tables would appear numerically similar. However, being different parameters, they are *not* even the same dimensionally. Thus the 'per hour' units are used to minimise confusion.

The reason for there being effectively two tables (high and low demand) is that there are two ways in which the integrity target may need to be described. The difference can best be understood by way of examples.

Consider the motor car brakes. It is the rate of failure which is of concern because there is a high probability of suffering the hazard immediately each failure occurs. Hence we have the middle column of Table 1.1.

On the other hand, consider the motor car air bag. This is a low demand protection system in the sense that demands on it are infrequent (years or tens of years apart). Failure rate alone is of little use to describe the integrity since the hazard is not incurred immediately each failure occurs and we therefore have to take into consideration the test interval. In other words, since the demand is infrequent, failures may well be dormant and persist during the test interval. What is of interest is the combination of failure rate and down time and we therefore specify the probability of failure on demand (PFD). Hence the right-hand column of Table 1.1.

In IEC 61508 the high demand definition is called for when the demand on a safety related function is greater than once per annum and the low demand definition when it is less frequent. There is some debate on this issue and it is believed that the classification might change. One possibility is that low demand might be defined as being when the demand rate is much less than the test frequency (i.e. reciprocal of the test interval).

In Chapter 2 we will explain the ways of establishing a target SIL and it will be seen that the IEC 61508 Standard then goes on to tackle the two areas of meeting the quantifiable target and addressing the qualitative requirements separately. Appendix 7 has more on the difference between the high and low demand scenarios.

A frequent misunderstanding is to assume that if the qualitative requirements of a particular SIL are observed the numerical failure targets, given in Table 1.1, will automatically be achieved. This is most certainly not the case since the two issues are quite separate. The quantitative targets refer to random hardware failures and are dealt with in Chapters 6–8. The qualitative requirements refer to quite different failures whose frequency is *not* quantified and are dealt with separately. The assumption, coarse as it is, is that by spreading the rigour of requirements across the range SIL 1–SIL 4, which in turn covers the credible range of achievable integrity, the achieved integrity is likely to coincide with the measures applied.

A question sometimes asked is:

If the quantitative target is met by the predicted random hardware failure probability then what allocation should there be for the systematic (software) failures? Note 1 of 7.4.2.2 of Part 2 of the Standard tells us that the target is to be applied equally to random hardware failures and to systematic failures. In other words the numerical target is not

divided between the two but applied to the random hardware failures. The corresponding SIL requirements are then applied to the systematic failures. In any case, having regard to the accuracy of quantitative predictions (see Chapter 7), the point may not be that important.

The following should be kept in mind:

SIL 1: is relatively easy to achieve especially if ISO 9001 practices apply throughout the design providing that Functional Safety Capability is demonstrated (see Section 2.1).
SIL 2: is not dramatically harder than SIL 1 to achieve although clearly involving more review and test and hence more cost. Again, if ISO 9001 practices apply throughout the design, it should not be difficult to achieve.

(SILs 1 and 2 are not dramatically different in terms of the life-cycle activities)

SIL 3: however, involves a significantly more substantial increment of effort and competence than is the case from SIL 1 to SIL 2. Specific examples are the need to revalidate the system following change and the increased need for training of operators. Cost and time will be a significant factor and the choice of vendors will be more limited by lack of ability to provide SIL 3 designs.
SIL 4: involves state of the art practices including 'formal methods' in design. Cost will be extremely high and competence in all the techniques required is not easy to find. There is a considerable body of opinion that SIL 4 should be avoided and that additional levels of protection should be preferred.

It is reasonable to say that the main difference between the SILs is the quantification of random hardware failures and he application of the Safe Failure Fraction (see Chapter 3). The qualitative requirements for SILs 1 and 2 are very similar, as are those for SILs 3 and 4. The major difference occurs in the step between SIL 2 and SIL 3.

Note, also, that as one moves up the SILs the statistical implications of verification become more onerous whereas the assessment becomes more subjective due to the limitations of the data available for the demonstration.

1.3 The life-cycle approach

The various life-cycle activities and defences against systematic failures, necessary to achieve functional safety, occur at different stages in the design and operating life of any equipment. Hence it has long been considered a good idea to define (that is to say describe) a life-cycle.

IEC 61508 describes itself as being based on a safety life-cycle approach and therefore it describes such a model and identifies activities and requirements based on it. It is important to understand this because a very large proportion of safety assessment work has been (and often still is) confined to assessing if the proposed design configuration (architecture) meets the target failure probabilities (Part C of this book). Most modern guidance (especially IEC 61508) requires a much wider approach involving control over all of the life-cycle activities that influence safety-integrity.

Figure 1.2 shows a simple life-cycle very similar to the one shown in the Standard. It has been simplified for the purposes of this book.

As far as IEC 61508 is concerned this life-cycle applies to all electrical and programmable aspects of the safety-related equipment. Therefore if a safety-related system contains an E/PE elem ent then the Standard applies to all the elements of system, including mechanical and pneumatic equipment. There is no reason, however, why it should not also be used in respect of 'other technologies' where they are used to provide risk reduction.

The IEC 61508 headings are summarised in the following pages and also map to the descriptions of many of the headings in Chapters 3, 4 and 5. This is because the Standard repeats the process for systems hardware (Part 2) and for software (Part 3). IEC 65108 Part 1 lists these and calls the list Table 1 with associated paragraphs of text. In the following text '*' refers to the IEC 61508 Part 1 Table. Also, the IEC 61508 paragraph numbers for the associated text, in Parts 1, 2 and 3, are given:

Life-cycle (*1) [Part 1 – 7.1/2: Part 2 – 7.1/2: Part 3 – 7.1/2]
Sets out the life-cycle for the development maybe as per IEC 61508, *or* as shown in Figure 1.2 of this book, *or* some other suitable format having regard to the project and to in-house practice.

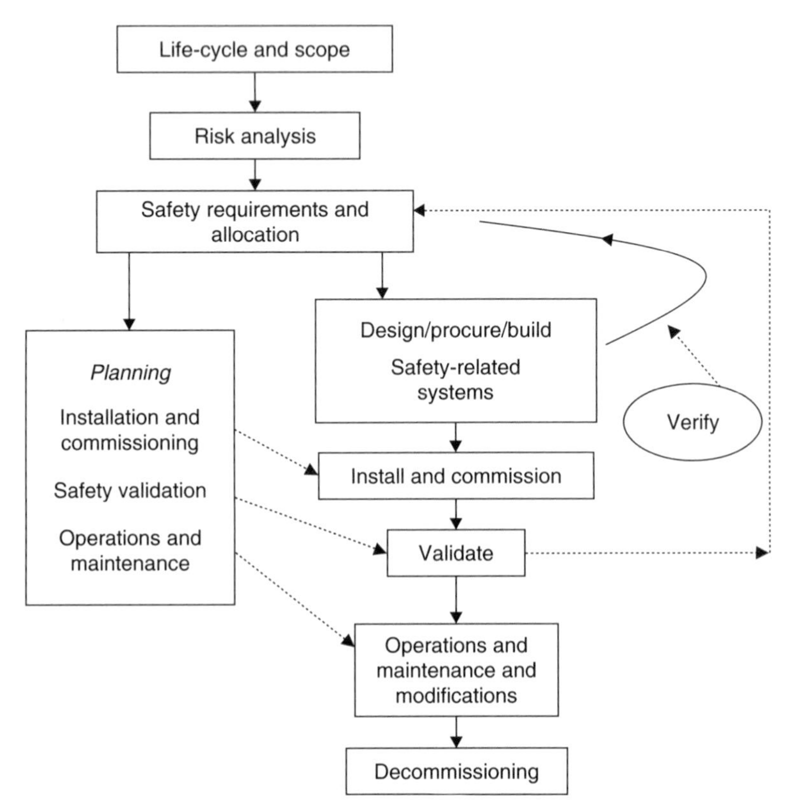

Figure 1.2
Safety life-cycle

Equipment Under Control (EUC) and scope (*2) [Part 1 – 7.3]
Defines exactly what is the system and the part(s) being controlled. Understands the EUC boundary and its safety requirements. Scopes the hazards and risks by means of hazard identification techniques (e.g. HAZOP). Requires a safety plan for all the life-cycle activities.

Hazard and risk analysis (*3) [Part 1 – 7.4]
This involves the quantified risk assessment by considering the consequences of failure (often referred to as HAZAN).

Safety requirements and allocation (*4/5) [Part 1 – 7.5/6: Part 2 – 7.2: Part 3 – 7.2]
Here we address the *whole system* and set maximum tolerable risk targets and allocate failure rate targets to the various failure

modes across the system. Effectively this defines what the safety function is by establishing what failures are protected against and how. Thus the safety functions are defined and *each* has its own SIL (see Chapter 2).

Plan operations and maintenance (*6) [Part 1 – 7.7: Part 2 – 7.6]
What happens in operations, and during maintenance, can effect functional safety and therefore this has to be planned. The effect of human error is important here as will be mentioned in Chapter 6. This also involves recording actual safety-related demands on systems as well as failures.

Plan installation and commissioning (*8) [Part 1 – 7.9]
What happens through installation and commissioning can effect functional safety and therefore this has to be planned. The effect of human error is important here as will be shown in Chapter 6.

Planning tests, operations etc. (i.e. validation) (*7) [Part 1 – 7.8: Part 2 – 7.3/4/5/7/9: Part 3 – 7.3/4]
It is necessary to plan ahead as to how reviews and tests will be structured. This is sometimes called a quality plan but often called validation planning. It includes integration and test specifications for hardware and software, test logs, reviews etc.

Design and build the system (*9–11) [Part 1 – 7.10 to 12: Part 2 – 7.4 to 8: Part 3 – 7.4 to 8]
This is called 'realisation' in IEC 61508. It means creating the actual safety systems be they electrical, electronic, pneumatic, or simply failure avoidance measures (e.g. physical bunds or barriers).

Install and commission (*12) [Part 1 – 7.13]
Implement the installation and create records of events during installation and commissioning, especially failures.

Validate that the safety-systems meet the requirements (*13) [Part 1 – 7.14: Part 2 – 7.5 and 7: Part 3 – 7.5 and 7]
This involves checking that all the allocated targets (above) have been met. This will involve a mixture of predictions, reviews and test results. There will have been a validation plan (*7 above) and there will need to be records that all the tests

have been carried out and recorded for both hardware and software to see that they meet the requirements of the target SIL. It is important that the system is revalidated from time to time during its life, based on recorded data.

Operate, maintain, and repair (*14) [Part 1 – 7.15: Part 2 – 7.6: Part 3 – 7.6]
Clearly operations and maintenance (already planned in *6 above) are important. Documentation, particularly of failures, is important.

Modifications (*15) [Part 1 – 7.16: Part 2 – 7.5/6/8: Part 3 – 7.8]
It is also important not to forget that modifications are, in effect, redesign and that the life-cycle activities should be activated as appropriate when changes are made.

Disposal (*16) [Part 1 – 7.17]
Finally, decommissioning carries its own safety hazards which should be addressed.

Verification (–) [Part 1 – 7.18: Part 2 – 7.9: Part 3 – 7.9]
Demonstrating that all life-cycle stage deliverables were met in use.

Assessments (–) [Part 1 – 8: Part 2 – 8: Part 3 – 8]
Carry out assessments to demonstrate compliance with the target SILs (see Chapter 2 for the extent of independence according to consequences and SIL).

1.4 Basic steps in the assessment process

The following steps are part of the safety life-cycle (already described). They are the parts referenced as (*3, *4 and *5) in Section 1.3 and refer to the risk and SIL assessment activities.

Step 1. Establish a risk target
ESTABLISH THE RISK TO BE ADDRESSED by means of techniques such as formal hazard identification or HAZOP whereby failures and deviations within a process (or equipment) are studied to assess outcomes. From this process one or more hazardous events may be revealed which will lead to death or serious injury.

SET MAXIMUM TOLERABLE RISK by carrying out some form of quantified risk assessment so that the probability of death or injury, arising from the event in question, is assessed. By considering the maximum tolerable risk (dealt with in the next chapter), and taking into account how many simultaneous risks to which one is exposed in the same place, a maximum tolerable failure rate for each event can be targeted.

Step 2. Identify the safety-related function
For each hazardous event it is necessary to understand what failure modes will lead to it. In this way the various elements of protection (e.g. control valve *and* relief valve *and* slamshut) can be identified. The safety protection system for which a SIL is needed can then be identified.

Step 3. Establish a target SIL for the safety-related element
The NUMERICAL ASSESSMENT and the RISK GRAPH methods are described in Chapter 2.

Step 4. Quantitative assessment of the safety-related system
Reliability modelling is needed to assess the failure rate or probability of failure on demand of the safety-related element or elements in question. This can then be compared with the target set in Step 3. Chapters 6–8 cover the main techniques.

Step 5. Qualitative assessment against the SILs
The various requirements for limiting systematic failures are more onerous as the SIL increases. These cover many of the life-cycle activities and are covered in Chapters 4 and 5.

Step 6. Establish ALARP
It is not sufficient to establish, in Step 4, that the quantitative failure rate (or the PFD) has been met. Design improvements which reduce the failure rate (until the Broadly Acceptable failure rate is met) should be considered and an assessment made as to whether these are 'as low as reasonably practicable'. This is covered in Section 2.3.

Step 7. Establish functional safety capability
Whereas the above steps refer to the assessment of a system or product, there is the additional requirement to establish the FUNCTIONAL SAFETY CAPABILITY of the assessor and/ or the design organisation. This is dealt with in Section 2.1 and in Appendix 1.

It is worth noting at this point that conformance to a SIL requires that all the STEPS are met. If the quantitative assessment (STEP 4) indicates a given SIL then this can only be claimed if the qualitative requirements (STEP 5) are also met.

1.5 Costs

The following questions are often asked:

- 'What is the cost of applying IEC 61508?'
- 'What are the potential savings arising from its use?' and
- 'What are the potential penalty costs of ignoring it?'

1.5.1 Costs of applying the Standard

Although costs will vary considerably, according to the scale and complexity of the system or project, the following typical resources have been seen in meeting various aspects of IEC 61508.

Full Functional Safety Capability (to the level of Accredited certification) including implementation on a project or product – 30 to 60 mandays + several £'000 for certification.

Product or Project Conformance (to the level of third party independent assessment) – 10–20 mandays + a few £'000 consultancy.

Elements within this can be identified as follows:

Typical SIL targeting with random hardware failures assessment and ALARP – two to six mandays

Assessing safe failure fraction (one or two failure modes) – one to five mandays

Bringing an ISO 9000 management system up to IEC 61508 functional safety capability – five mandays for the purpose of a product demonstration, 20 to 50 mandays for the purpose of an accredited certificate.

1.5.2 Savings

There is an intangible but definite benefit due to enhanced credibility in the market place. Additional sales vis-à-vis those who have not demonstrated conformance are likely.

Major savings are purported to be made in reduced maintenance for those (often the majority of) systems which are given low SIL targets. This also has the effect of focusing the effort on the systems with higher SIL targets.

1.5.3 Penalty costs

The manufacturer and the user will be involved in far higher costs of retrospective redesign if subsequent changes are needed to meet the maximum tolerable risk.

The user could face enormous legal costs in the event of a major incident which invokes the H&SW Act especially if the Standard had not been applied when it was reasonably practicable to do so.

1.6 The seven parts of IEC 61508

Now that we have introduced the two ideas of safety-integrity levels and a life-cycle approach it is now appropriate to describe the structure of the IEC 61508 Standard. Parts 1–3 are the main parts and Parts 4–7 provide supplementary material.

The general strategy is to establish SIL targets, from hazard and risk analysis activities, and then to design the safety-related equipment to an appropriate integrity level taking into account random and systematic failures and also human error.

Examples of safety-related equipment might include:

Shutdown systems for processes
Interlocks for dangerous machinery
Fire and gas detection
Railway signalling
Boiler and burner controls
Leisure items (e.g. fairground rides)
Medical equipment (e.g. oncology systems)

Part 1 is called 'General requirements'. It covers:

(i) General functional safety management, dealt with in Chapter 2 of this book. This is the management system (possibly described in one's quality management system) which lays down the activities, procedures and skills necessary to carry out the business of risk assessment and of designing to meet integrity levels.

(ii) The life-cycle, explained above, and the requirements at each stage, are central to the theme of achieving functional safety. It will dominate the structure of several of the following chapters and appendices.

(iii) The definition of SILs and the need for a hazard analysis in order to define a SIL target.

(iv) The need for competency criteria for people engaged in safety-related work, also dealt with in Chapter 2 of this book.

(v) Levels of independence of those carrying out the assessment. The higher the SIL the more independent should be the assessment.

Chapter 2 is devoted to summarising Part 1 of IEC 61508.

(i) There is an annex in Part 1 (informative only) providing a sample document structure for a safety-related design project.

(ii) There is also an annex listing factors relevant to competency which will also be dealt with in Chapter 2.

Part 2 is called 'Requirements for E/E/PES safety-related systems'. What this actually means is that Part 2 is concerned with the hardware aspects of the safety-related system, rather than the software. It covers:

(i) The life-cycle activities associated with the design and realisation of the equipment including defining safety requirements, planning the design, validation, verification, observing architectural constraints, fault tolerance, test, subsequent modification (all of which will be dealt with in Chapter 3).

(ii) The need to assess (i.e. predict) the quantitative reliability (vis-à-vis random hardware failures) against the SIL targets in Table 1.1. This is the reliability prediction part of the process and is covered in Chapters 6 and 7.

(iii) The techniques and procedures for defending against systematic hardware failures.

(iv) Architectural constraints vis-à-vis the amount of redundancy applicable to each SIL. Hence, even if the above reliability prediction indicates that the SIL is met, there will still be minimum levels of redundancy. This could be argued as being because the reliability prediction will only have addressed random hardware failures (in other words those present in the failure rate data) and there is still the need for minimum defences to tackle the systematic failures.

(v) Some of the material is in the form of annexes which are informative.

Chapter 3 is devoted to summarising Part 2 of IEC 61508.

Part 3 is called 'Software requirements'. As the title suggests this addresses the activities and design techniques called for in the design of the software. It is therefore about systematic failures and no quantitative prediction is involved.

(i) Tables indicate the applicability and need for various techniques at each of the SILs.
(ii) Some of the material is in the form of annexes which are informative.

Chapter 4 is devoted to summarising Part 3 of IEC 61508.

Part 4 is called 'Definitions and abbreviations'. This book does not propose to offer yet another list of terms and abbreviations beyond the few terms in Appendix 8. In this book the terms are hopefully made clear as they are introduced.

Part 5 is called 'Examples of methods for the determination of safety-integrity levels'. As mentioned above, the majority of Part 5 is in the form of five Annexes which are informative rather than normative:

(i) Annex A covers the general concept of the need for risk reduction through to the allocation of safety requirements, which is covered in Chapter 2 of this book.
(ii) Annex B covers the application of the ALARP (as low as reasonably practicable) principle, which is covered in Chapter 2 of this book.
(iii) Annex C covers the mechanics of quantitatively determining the SIL levels, which is covered in Chapter 2 of this book.

(iv) Annex D covers a qualitative method (risk graph) of establishing the SIL levels, which is also covered in Chapter 2 of this book.

(v) Annex E describes an alternative qualitative method, 'Hazardous event severity matrix', which is not too dissimilar to the one described at the end of Chapter 2.

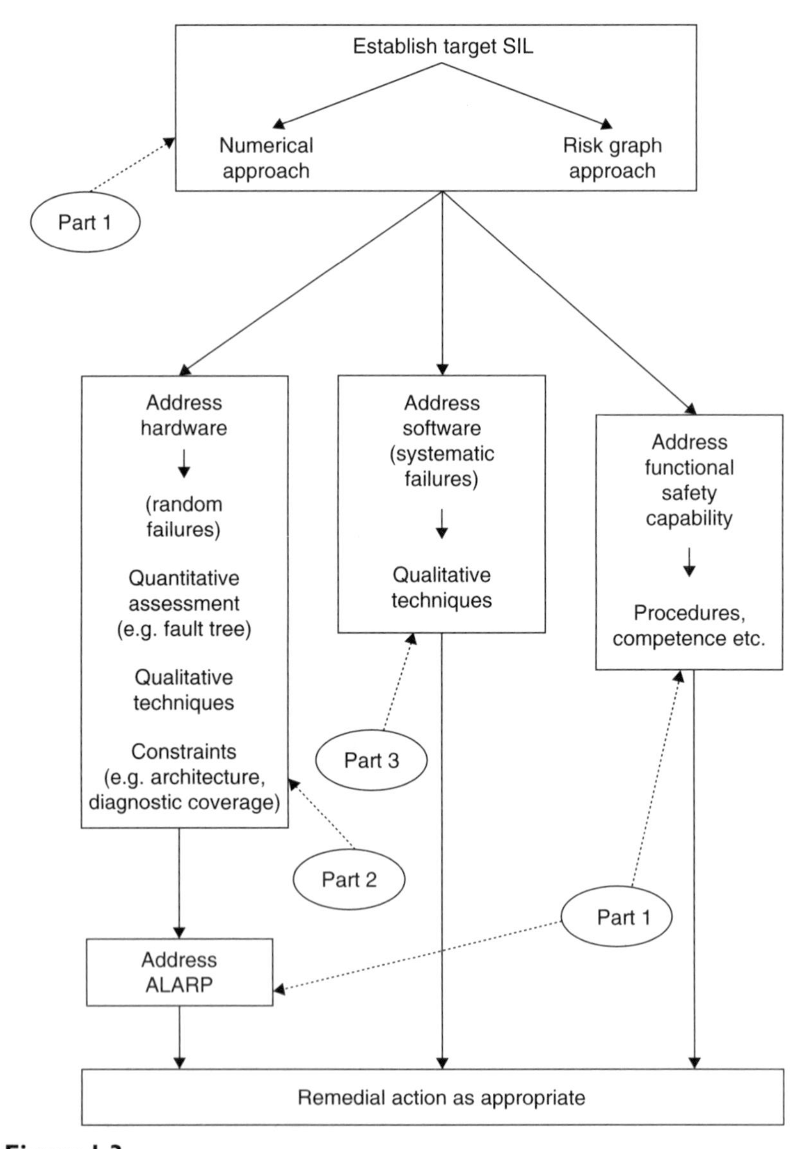

Figure 1.3
The parts of the Standard

Part 6 is called 'Guidelines on the application of Part 2 and Part 3'. This consists largely of informative annexes which provide material on:

(i) Calculating hardware failure probabilities (low and high demand), which is covered in Chapter 8 of this book.
(ii) Common cause failure, which is covered in Chapter 6 of this book.
(iii) Diagnostic coverage, which is covered in Chapter 3 of this book.
(iv) Applying the software requirements tables (of Part 3) for SILs 2 and 3, which is covered in Chapter 4 of this book.

As mentioned above, the majority of Part 6 is in the form of Annexes which are 'informative' rather than 'normative'.

Part 7 is called 'Overview of techniques and measures'. This is a reference guide to techniques and measures and is cross-referenced from other parts of the Standard. This book does not repeat that list but attempts to explain the essentials as it goes along.

The contents of Parts 1–2 of the Standard are illustrated diagrammatically in Figure 1.3 and the requirements summarised in Figure 1.4.

Targeting SILs
Assessing random hardware failures
Meeting ALARP
Assessing architectures
Meeting the life-cycle requirements
Having the functional capability to achieve the above

Figure 1.4
Summary of the requirements

PART B

THE BASIC REQUIREMENTS OF IEC 61508 AND 61511

In this section Chapters 2 to 5 will summarise the requirements of:

IEC 61508 Part 1
IEC 61508 Part 2
IEC 61508 Part 3
IEC 61511

CHAPTER 2

MEETING IEC 61508 PART 1

Part 1 of the Standard covers the need for:

- Capability to design, operations and maintenance for functional safety
- Establishing competency
- Setting SIL targets
- The ALARP concept

The following Sections summarise the main requirements.

2.1 Functional safety management and competence

2.1.1 Functional Safety Capability assessment

In claiming conformance to any of the SILs it is necessary to show that the management of the design, operations and maintenance activities and of the system implementation is itself appropriate and that there is adequate competence for carrying out each task.

This involves two basic types of assessment. The first is the assessment of management procedures (very similar to an ISO 9000 audit). Appendix 1 of this book provides a Functional Safety Capability template procedure which should be adequate as an addition to an ISO 9000 quality management system. The second is an assessment of the implementation of these procedures. Thus, the life-cycle activities described in

Chapters 1, 3, 4 and 5 would be audited, for one or more projects, to establish that the procedures are being put into practice.

Appendix 2 contains a checklist schedule to assist in the rigour of assessment, particularly for self assessment (see also Chapter 10.3).

2.1.2 Competency

In Part 1 of IEC 61508 (Paragraph 8.2.11 and Annex B) the need for adequate competency is called for. Annex B is open ended in that it only calls for the training, knowledge, experience and qualifications to be 'relevant'. Factors listed for consideration are:

- Engineering application knowledge
- Technology knowledge
- Safety engineering knowledge
- Legal/regulatory knowledge
- The link between magnitude of consequences and rigour of competence
- The link between SIL and rigour of competence
- The link between design novelty and rigour of competence
- Relevance of previous experience
- Relevance of qualifications
- The need for training to be documented

A much quoted guidance document in this area is the IEE/BCS (Institution of Electrical Engineers and British Computer Society) document 'Competency Guidelines for Safety-related Systems Practitioners'. In this, 12 safety-related job functions (described as functions) are identified and broken down into specific tasks. Guidance is then provided on setting up a review process and in assessing capability (having regard to applications relevance) against the interpretations given in the document. The 12 jobs are:

1. *Corporate Functional Safety Management:* This is relevant to the Functional Safety Capability requirement described in Appendix 1 of this book. It concerns the competency required to develop and administer this function within an organisation.

2. *Project Safety Assurance Management:* This extends the previous task into implementing the functional safety requirements in a project.
3. *Safety-Related System Maintenance:* This involves maintaining a system and controlling modifications so as to maintain the safety-integrity targets.
4. *Safety-Related System Procurement:* This covers the technical aspects of controlling procurement and sub-contracts (not just administration).
5. *Independent Safety Assessment:* This is supervising and/or carrying out the assessments.
6. *Safety Hazard and Risk Analysis:* That is to say HAZOP (HAZard and OPerability study), risk analysis, prediction etc.
7. *Safety Requirements Specification:* Being able to specify all the safety requirements for a system.
8. *Safety Validation:* Defining a test/validation plan, executing and assessing the results of tests.
9. *Safety-Related System Architectural Design:* Being able to partition requirements into sub-systems so that the overall system meets the safety targets.
10. *Safety-Related System Hardware Realisation:* Specifying hardware and its tests.
11. *Safety-Related System Software Realisation:* Specifying software, developing code and testing the software.
12. *Human Factors Safety Engineering:* Assessing human error and engineering the inter-relationships of the design with the human factors (Chapter 6.4).

The three levels of competence described in the document are:

1. *The Supervised Practitioner* who can carry out one of the above jobs but requiring review of the work.
2. *The Practitioner* who can work unsupervised and can manage and check the work of a Supervised Practitioner.
3. *The Expert* who will be keeping abreast of the state of art and will be able to tackle novel scenarios.

Tables are provided for each of the 12 functions described above. The function is described and FUNCTION related competencies with guidance as to what describes a Supervised Practitioner, Practitioner or Expert.

An example (extract) for the Task: ASSESSING SAFETY ANALYSIS within the INDEPENDENT SAFETY ASSESSMENT JOB is:

Supervised practitioner	*Practitioner*	*Expert*
… performed activities requiring the use of relevant analysis techniques and can illustrate this with …	Can illustrate how relevant techniques (e.g. fault trees) have been used … support a conclusion …	… illustrate how inappropriate techniques can lead to unsafe conclusions …

Name:	Mr A.N. Other
Job Title:	Technical Manager
Date of Birth:	01-03-1950
Qualifications:	BSc, MSaRS, MSc (Safety and Reliability) XY University
Date of XYZ Employment:	Project EXXX
Application Domain Knowledge:	Sectors – Oil and gas only Project EXXX (Morecambe Bay) SR code in 'C' (Sil 2 – one year's experience) Project EXXX (???) SR code in GE application language (SIL 1 – one year's experience)
Accuracy and Detail:	Project EXXX
Decisions/Communication/ Inter-working:	Good (Ref. 02-01-2001 appraisal)
FS Assurance:	No
Functional Safety and Regulatory Knowledge:	Attended in-house course (31-01-88) Knows IEC 61508 and has reviewed draft 61511 Part 1 with MD
Testing:	Participated in the FS testing of Project EXXX
Reviews:	No
FS Audits:	Reviewed the EXXX FS audit
Bidding for Work	No
Safety Authority:	Project EXXX 1999 (SIL 2)
Assessing Individuals on this Register:	No

Figure 2.1
Competency assessment

It is intended that these guidelines be updated and improved with use.

This IEE/BCS document provides a solid basis for the development of competence. It probably goes beyond what is actually called for in IEC 61508. Due to its complexity it is generally difficult to put into practice in full and therefore might discourage some people from starting a scheme. Hence a simpler approach might be more practical. However, this is a steadily developing field and the requirements of 'good practice' are moving forward.

The minimum requirement in this area should surely be:

1. A documented statement (for each safety-related job) of knowledge and skill.
2. Documented evidence of objective review of all individuals involved in the safety life-cycle.

Figure 2.1 shows a typical interpretation of the need for an Assessment Document for each person.

2.1.3 Independence of the assessment

Throughout the life-cycle the level of independence to be applied when carrying out assessments is recommended, according to SIL, as:

SIL	*Assessed by:*
4	Independent organisation
3	Independent department
2	Independent person
1	Independent person

For SILs 2 and 3 add one level of independence if there is lack of experience, unusual complexity or novelty of design. Clearly, these terms are open to interpretation and words such as 'department' and 'organisation' will depend on the size and type of company. For example, in a large multi-project design company there might be a separate safety assessment department sufficient to meet the requirements of SIL 3. A smaller single-project company might, on the other hand, need to engage an independent organisation or consultant in order to meet the SIL 3 requirement.

The level of independence to be applied when establishing SIL targets is recommended, according to consequence, as:

Multiple fatality say >5	Independent organisation
Multiple fatality	Independent department
Single fatality	Independent person
Injury	Independent person

For scenarios involving fatality, add one level of independence if there is lack of experience, unusual complexity or novelty of design. Clearly, these terms are open to interpretation and words such as 'department' and 'organisation' will depend on the size and type of company.

2.2 Establishing SIL targets

Assessing the amount of risk reduction required from a protection system is an essential part of the design process. The following paragraphs describe how a SIL target is established.

2.2.1 Quantitative approach

(a) As an example of selecting an appropriate SIL, assume that the maximum tolerable frequency for an involuntary risk scenario (e.g. customer killed by explosion) is 10^{-5} pa (A) (see Table 2.1). Assume that 10^{-2} (B) of the hazardous events in question lead to fatality. Thus the maximum tolerable failure rate for the hazardous event can be $C = A/B = 10^{-3}$ pa. Assume a fault tree analysis indicates that the unprotected process is only likely to achieve a failure rate of 2×10^{-1} pa (D) (i.e. 1/5 years). The FAILURE ON DEMAND of the safety system would need to be $E = C/D = 5 \times 10^{-3}$. Consulting the right-hand column of Table 1.1, SIL 2 is applicable.

This is an example of a **low demand** safety-related system in that it is only called upon to operate at a frequency determined by the frequency of failure of the equipment under control (EUC) – in this case 2×10^{-1} pa. Note also that the target 'E' in the above paragraph is dimensionless by virtue of dividing a rate by a rate. Again, this is consistent with the right-hand column of Table 1.1 in Chapter 1.

(b) Now consider an example where a failure in a domestic appliance leads to overheating and subsequent fire. Assume, again, that the target risk of fatality is said to be 10^{-5}pa. Assume that a study suggests that 1 in 400 incidents leads to fatality.

It follows that the target maximum tolerable failure rate for the hazardous event can be calculated as $10^{-5} \times 400 = 4 \times 10^{-3}$pa (i.e. 1/250 years). This is 4.6×10^{-7}per hour when expressed in units of 'per hour' for the purpose of Table 1.1.

Consulting the middle column of Table 1.1, SIL 2 is applicable. This is an example of a **high demand** safety-related system in that it is 'at risk' continuously. Note also that the target in the above paragraph has the dimension of rate by virtue of multiplying a rate by a dimensionless number. Again, this is consistent with the middle column of Table 1.1.

It is worth noting that for a low demand system the Standard, in general, is being applied to an 'add-on' safety system which is separate from the normal control of the EUC (i.e. plant). On the other hand, for a continuous system the Standard, in general, is being applied to the actual control element because its failure will lead directly to the potential hazard even though the control element may require additional features to meet the required integrity.

The above two examples imply a need for a maximum tolerable risk target. This is a controversial area dealt with at length in the HSE documents 'TOR' and 'R2P2' (see Appendix 6). However, typical figures such as those mooted in IGEM SR/15 (Chapter 9) are shown in Table 2.1.

Table 2.1 Target risk levels

Maximum tolerable risk of fatality	*Individual risk (per annum)*
Employee	10^{-4}
Public	10^{-5}
Broadly acceptable risk (previously referred to as 'Negligible' (Employee and public))	10^{-6}

The table should be viewed in the light of the following information:

All accidents (per individual)	5×10^{-4} pa
Natural disasters (per individual)	2×10^{-6} pa
Accident in the home	4×10^{-4} pa
Worst case maximum tolerable risk in HSE R2P2 document*	10^{-3} pa
'Very low risk' as described in HSE R2P2 document (i.e. boundary between Tolerable and Broadly Acceptable)	10^{-6} pa

*Table 2.1 is one order more rigorous for the boundary of unacceptable and ALARP risk.

It is interesting to note that a figure of 10^{-6} pa for greater than five fatalities arising from the storage of nuclear waste has been quoted in the press. The meaning and significance of BROADLY ACCEPTABLE will be dealt with when explaining ALARP in Section 2.3.

The number of sources of risk need to be kept in mind when applying these ideas. For example, a target maximum tolerable risk of 10^{-5} pa may have been chosen for the assessment of a risk from a process. There may, however, be ten of those processes, or nine other similar events from the one process, all within the same vicinity and capable of affecting the same person (*in the same place*). In that case the maximum tolerable risk for each event being assessed might well be adjusted to 10^{-6} pa.

Another less quantified approach is by means of subjective risk severity classification described at the end of this chapter.

Now try the following exercises (answers in Appendix 5) which involve establishing SIL targets:

Exercise 1:

Assume a maximum tolerable risk target of 10^{-5} pa (Public fatality).

Assume 1 in 10 incidents lead to a fatality.

Assume that a fault tree indicates that the process will suffer a failure rate of 0.05 pa.

It is proposed to implement an add-on safety system involving instrumentation and shutdown measures.

Which type of SIL is indicated and why?
Which SIL level should be the target?

Exercise 2:
2.1
Assume a maximum tolerable risk fatality target of 10^{-5} pa.
Assume that there are nine other similar hazards to be assessed from the plant which will threaten the same group of people at the same time.
Assume spillage causes fatality 1 in 10 times.
Assume that a fault tree indicates that each of the processes will suffer an incident once in 50 years.
It is proposed to implement an add-on safety system with instrumentation and shutdown measures
Which type of SIL is indicated and why?
Which SIL level should be the target?

2.2
If additional fire fighting equipment were made available, to reduce the likelihood of a fatality from 1 in 10 to 1 in 30, what effect, if any, is there on the target SIL?

These involved the low demand table in which the risk criteria were expressed as a probability of failure on demand (PFD). Now try Exercise 3.

Exercise 3:
Target maximum tolerable risk = 10^{-5} pa.
Assume that 1 in 200 failures, whereby an interruptible gas meter spuriously closes and then opens, leads to fatality.
Which type of SIL is indicated and why?
Which SIL level should be the target?

A point worth pondering is that when a high demand SR system fails continued use is usually impossible; for the low demand system limited operation may still be feasible after the risk reduction system has failed, albeit with additional care.

Appendix 7 provides some additional comparison of the low and high demand cases.

A methodology, specifically mentioned in Part 3 of IEC 61511 (Annex F), is known as Layer of Protection Analysis (LOPA).

Essentially, it is the same quantitative approach that has been described in the foregoing examples. However, it has the advantage of defining specific items to be addressed, as, for example:

1. Initiation likelihood (for the scenario)
2. Types of mitigation (pressure relief, bunds, deluge etc.)
3. Corporate risk criteria (i.e. maximum tolerable risk)

This provides a useful checklist particularly for the non-safety specialist. As with all checklists the point worth emphasising is that they are very useful providing that they are not allowed to constrain.

2.2.2 The risk graph approach

The IEC Standard acknowledges that a quantitative approach to setting SIL targets is not always possible and that an alternative approach might sometimes be appropriate. This avoids quantifying the maximum tolerable risk of fatality and uses qualitative judgements. Figure 2.2 gives an example of a risk graph as used in the UKOOA guidelines for the offshore oil and gas industry (see Chapter 9).

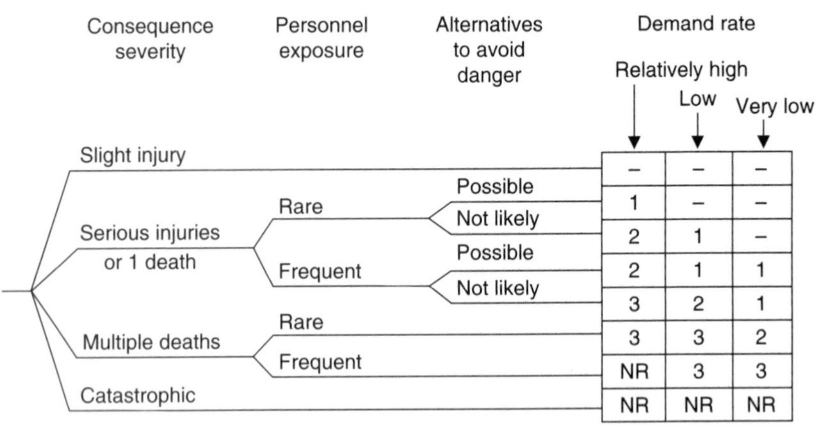

– = No special safety features required
NR = Not recommended. Consider alternatives

Figure 2.2
Risk graph

Note that in Figure 2.2 'Relatively high' = 3 pa, 'Low' = 0.3 pa, 'Very Low' = 0.03 pa. It has been suggested that in practice there is a need for a lower demand rate category ('Very very low'?) which might be 0.003 pa.

Although an advantage of this approach is that the risk graph is easier and quicker to apply, it is less precise. Interpretations of terms such as 'rare', 'possible' etc. can vary between assessors. There is therefore the need to calibrate the graph and to give guidance on the meanings of terms (e.g. rare). Without quantification, this is not easy since the SILs are defined in numerical terms.

A further point to consider is that, in view of the structure of the graph (Figure 2.2) wherein there is a 'demand rate' parameter to be specified, the risk graph approach therefore only lends itself to the low demand case.

It should also be noted that the QRA approach described in Section 2.2.1 targets a SIL based on Individual Risk and is not affected by the number of fatalities which are accounted for later in the ALARP calculation (see Section 2.3). The majority of risk graphs, however, take the number of fatalities into account when targeting the SIL. This is not consistent with the QRA approach. This is a particular problem in the process sector when trying to apply the risk graph approach for fire and gas systems where the number of potential fatalities exceeds one. Thus the risk graph might lead to a pessimistic SIL target.

Figure 2.3 shows a typical calibrated risk graph which takes account of injury as well as fatality in establishing the parameter shown as 'C'.

Now try Exercise 4.

Exercise 4:
Given that a single operator (at risk) is usually in attendance and has little opportunity for speedy escape or alternative mitigation, and that there is a moderate release of hydrocarbon, then:

Repeat Exercise 1 using the risk graph in Figure 2.3.

Risk graph approaches will not always be appropriate but they can be useful for screening as a means of quickly assigning priorities. Thus if the risk graph method suggests that nine

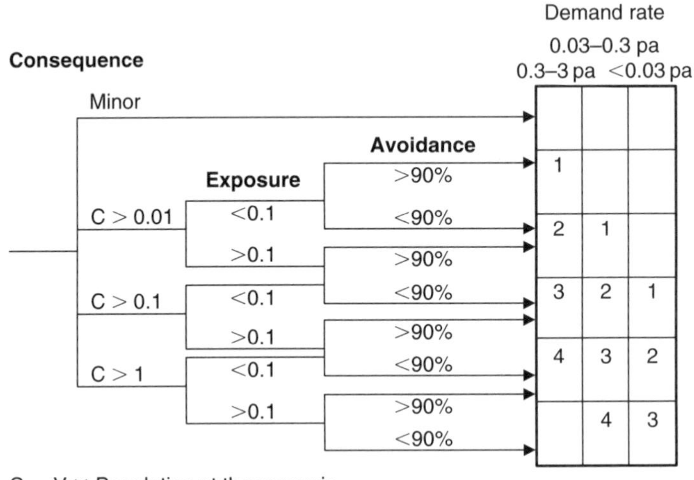

Figure 2.3

safety functions have a target of SIL 1 and only one has a target of SIL 3 then attention might well be directed to quantifying the latter. This tends to assume that the risk graph has given a pessimistic result and that it is 'safe' to accept a SIL 1 indication from it. This may not, of course, be the case.

> **IMPORTANT: It should be clear from the foregoing Sections that SILs are ONLY appropriate to specifically defined safety functions. A safety function might consist of a flow transmitter, logic element and a solenoid valve to protect against high flow. The flow transmitter, on its own, does not have a SIL and to suggest such is nearly meaningless. Its target SIL may vary from one application to another. The only way in which it can claim any SIL status in its own right is in respect of the life-cycle activities during its design, and this will be dealt with in Chapters 3, 4 and 5.**

2.2.3 'Not safety-related'

It may be the case that the SIL assessment indicates a probability of failure less than is indicated for SIL 1. In this case the system may be described as 'NOT SAFETY-RELATED' in the sense of the Standard. However, since the qualitative requirements of SIL 1 are little more than established engineering practice they should be regarded as a 'good practice' target.

The following example shows how a piece of control equipment might be justified to be 'NOT SAFETY-RELATED'. Assume that this programmable Distributed Control System (say a DCS for a process plant) causes various process shutdown functions to occur. In addition, let there be a hardwired Emergency Shutdown (presumably safety-related) system which can also independently bring about these shutdown conditions.

Assume the target maximum tolerable risk leads us to calculate that the failure rate for the DCS/ESD combined should be better than 10^{-3}pa. Assessment of the emergency shutdown system shows that it will fail with a PFD of 5×10^{-3}. Thus, the target failure rate of the DCS becomes 10^{-3}pa/$5 \times 10^{-3} = 2 \times 10^{-1}$pa. This being less onerous than the target for SIL 1 the target for the DCS is less than SIL 1. This is ambiguously referred to as 'not safety-related'. An alternative term used in some guidance documents is 'no special safety requirement'.

We are therefore justified in saying that the DCS is not safety-related. If, on the other hand, the target was only met by a combination of the DCS and ESD then each might be safety-related with a SIL appropriate to its target PFD or failure rate.

2.2.4 Environment and loss of production

So far the implication has been that integrity is in respect of failures leading to death or injury. IEC 61508 (and some other guidance) also refers to severe environmental damage. The UKOOA guidance (Chapter 9) provides a risk graph for establishing a SIL for equipment where failure leads to such an outcome (Figure 2.4). It is not known how the Figure 2.4 algorithm was developed.

Furthermore, although not directly relevant here, the same SIL approach can be applied to loss of production and, again, the UKOOA document provides a risk graph approach.

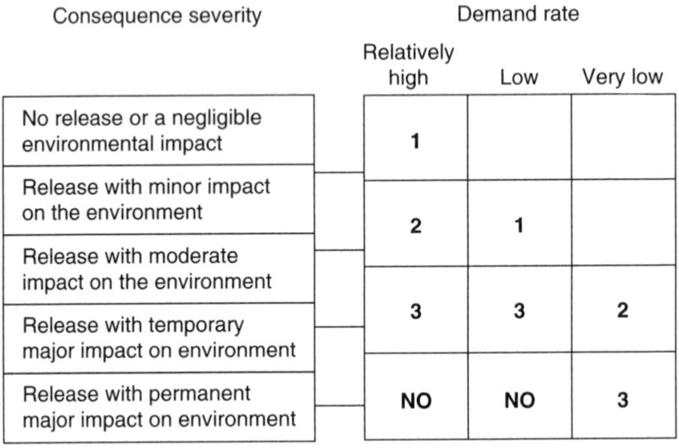

Consequence severity

- No release or a negligible environmental impact
- Release with minor impact on the environment
- Release with moderate impact on the environment
- Release with temporary major impact on environment
- Release with permanent major impact on environment

Demand rate: Relatively high	Low	Very low
1		
2	1	
3	3	2
NO	NO	3

Figure 2.4
Environmental risk graph

An alternative approach would be to establish a 'maximum acceptable annual cost'. Then, the probability of failure on demand might be assessed as the ratio:

$$\frac{\text{'Maximum acceptable annual cost'}}{(\text{Cost of the consequence} \times \text{Frequency of occurrence})}$$

The PFD could then be translated into a SIL using the low demand table.

2.3 Applying ALARP

The above section showed how a SIL target can be established either from a fatality target or by means of a risk graph.

Having established a SIL target it is not sufficient merely to assess that the design will meet the maximum tolerable risk target. It is necessary to establish if improvements are justified and thus the principle of ALARP (as low as reasonably practicable) is called for. This is implied by paragraphs 7.4.2.9 and 10 of IEC 61508 Part 1 which refer to 'good practice'. In the UK this is also necessary in order to meet safety legislation.

Figure 2.5 shows the so-called ALARP triangle which makes use of the idea of a Maximum Tolerable Risk.

In this context 'acceptable' is generally taken to mean that we accept the probability of fatality as being reasonably low,

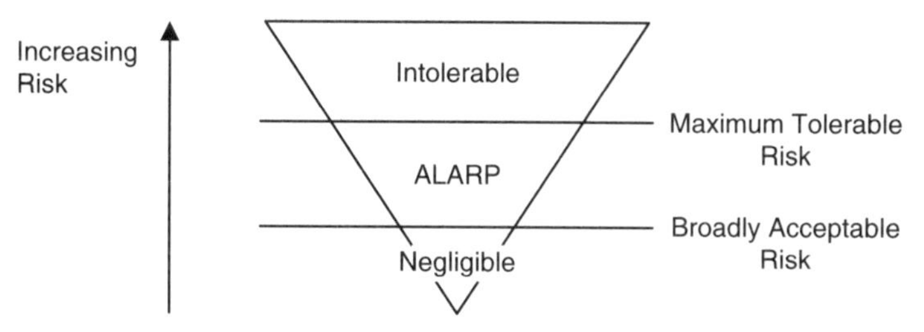

Figure 2.5
ALARP triangle

having regard to the circumstances, and would not usually seek to expend more resources in reducing it further.

'Tolerable', on the other hand, implies that whilst we are prepared to live with the particular risk level we would continue to review its causes and the defences we might take with a view to reducing it further. Cost comes into the picture in that any potential reduction in risk would be compared with the cost needed to achieve it.

'Unacceptable' means that we would not normally tolerate that level of risk and would not participate in the activity in question nor permit others to operate a process that exhibited it except perhaps in exceptional circumstances.

The principle of ALARP (as low as reasonably practicable) describes the way in which risk is treated legally and by the HSE in the UK, and also applied in some other countries. The concept is that all reasonable measures will be taken in respect of risks which lie in the tolerable (ALARP) zone to reduce them further until the cost of further risk reduction is grossly disproportionate to the benefit.

It is at this point that the concept of 'cost per life saved' arises. Industries and organisations are reluctant to state specific levels of 'cost per life saved' which they would regard as being grossly disproportionate to a reduction in risk. Figures in the range £1 000 000 to £15 000 000 are not infrequently quoted.

Perception of risk is certainly influenced by the circumstances. A far higher risk is tolerated from voluntary activities than involuntary ones (people feel that they are more in control of the situation on roads than on a railway). This explains the use of different targets for employee (voluntary) and public (involuntary) in Table 2.1.

A typical ALARP calculation might be as follows:

A £1 000 000 cost per life saved target is used in a particular industry.

A maximum tolerable risk target of 10^{-4} pa has been set for a particular hazard which is likely to cause two fatalities.

The proposed system has been assessed and a predicted risk of 8×10^{-5} pa obtained.

Given that the negligible risk is taken as 10^{-6} pa then the application of ALARP is required.

For a cost of £3000, additional instrumentation and redundancy will reduce the risk to just above the negligible region (2×10^{-6} pa).

The plant life is 30 years.

Hence cost per life saved = £3000/($8 \times 10^{-5} - 2 \times 10^{-6}$) $\times$ 2×30 = £640 000

This being less than the £1 000 000 cost per life saved criterion the proposal should be adopted. It should be noted that all the financial benefits of the proposed risk reduction measures should be included in the cost benefit calculation (e.g. saving plant damage, loss of production, business interruption etc.). Furthermore, following 'good practice' is also important although not of itself sufficient to demonstrate ALARP. Cost benefit arguments should not be used to justify circumventing established good practice.

Exercise 5:

A £2 000 000 cost per life saved target is used in a particular industry.

A maximum tolerable risk target of 10^{-5} pa has been set for a particular hazard which is likely to cause three fatalities.

The proposed system has been assessed and a predicted risk of 8×10^{-6} pa obtained.

How much could justifiably be spent on additional instrumentation and redundancy to reduce the risk from 8×10^{-6} pa to 2×10^{-6} pa (just above the negligible region).

The plant life is 25 years.

An alternative approach to ALARP, described in IEC 61508 Part 5, is known as the Risk Classification approach. It involves subjective assessments of Consequence and Frequency as, for example, in the following:

FREQUENCY	*CONSEQUENCE*			
	CATASTROPHIC	*CRITICAL*	*MARGINAL*	*NEGLIGIBLE*
FREQUENT	1	1	1	2
PROBABLE	1	1	2	3
OCCASIONAL	1	2	3	3
REMOTE	2	3	3	4
IMPROBABLE	3	3	4	4
INCREDIBLE	4	4	4	4

The Frequency/Consequence combination is then interpreted against the ALARP triangle as, for example:

INTOLERABLE REGION	1
TOLERABLE (ALARP) REGION	2 and 3
BROADLY ACCEPTABLE REGION	4

This method does not, however, lend itself immediately to the ALARP approach to justifying proposed modifications (implying a cost per life saved criteria), as described above. It does, however, give a quick indication of which region the scenario represents. In the event of its being in the tolerable (ALARP) region (classifications 2 and 3) then further justification is required to show that the risks usually are ALARP, maybe using a combination, of approaches. These might include the quantification described earlier in this Chapter as well as demonstrating that approved codes of practice have been followed.

Chapter 3

Meeting IEC 61508 Part 2

IEC 61508 Part 2 covers the safety system hardware and overall system design, whereas software design is covered by Part 3 (see next chapter). Sections 3.1–3.9 summarise the main requirements. However, the following points should be noted first.

The degree of variability which is used in the specification of techniques and measures is far greater than could ever be reasonably correlated with actual performance.

In any case each technique and the degree of refinement (e.g. high medium low) arises from an individual opinion of someone involved in the drafting process.

The combination of text (e.g. paras 7.1 to 7.9) and tables (both A and B series) and the use of modifying terms (such as high, medium and low) to describe the intensity of each technique has led to a highly complex set of requirements. Their interpretation requires the simultaneous reading of textual paragraphs, A tables, B tables and Table B6 – all on different pages of the Standard. The A tables are described as referring to measures for controlling (i.e. revealing) failures and the B tables to avoidance measures.

The authors of this book have, therefore, attempted to simplify this 'algorithm of requirements' and this chapter is offered as a credible representation of requirements for the four SILs.

At the end of this chapter a 'conformance demonstration template' is suggested which, when completed for a specific product or system assessment, will provide evidence of conformance to the SIL in question.

The approach to the assessment will differ substantially between:

COMPONENT (e.g. Transducer) DESIGN (CASS Type 1 (see Chapter 10))
and
APPLICATIONS SYSTEM DESIGN (CASS Type 2 (see Chapter 10))

The demonstration template tables at the end of this chapter cater for the latter case (i.e. Type 2). Chapter 5, involving the restricted subset of IEC 61511, also caters for the Type 2 case.

3.1 Organising and managing the life-cycle

Sections 7.1 and 7.3 of the Standard – Table '1'

The idea of a design life-cycle has already been introduced to cover all the activities during design, manufacture, installation and so on. The exact nature of the design-cycle model will depend on complexity and the type of system being designed. The IEC 61508 model (in Part 1 of the Standard) may well be suitable and the model in Chapter 1 of this book is very similar.

A major point worth making is that the life-cycle activities should all be documented. Unless this is done, there is no visibility to the design process and an assessment cannot verify that the standard has been followed. This should be a familiar discipline in as much as most readers will be operating within an ISO 9001 standard of practice. The design should be conducted under a project management regime and adequately documented to provide traceability. These requirements can be met by following a quality system such as specified in ISO 9001. The level and depth of the required project management and documentation will depend on the SIL level. The use of checklists is desirable at all stages.

The need for Functional Safety Capability has been described in Chapter 2, Section 2.1. IEC 61508 Part 2 (as well as Part 3 for the software) expects this to have been addressed.

In IEC 61508 Part 2 its Table '1' describes the life-cycle activities again and is more or less a repeat of Part 1.

Irrespective of SIL there needs to be a basic project management structure which defines all the required actions and responsibilities, along with defining adequate competency, of the personnel responsible for each task. There needs to be a 'Quality and Safety' plan which heads up the documentation hierarchy and describes the overall functional safety targets and plans. All documentation and procedures need to be well structured, for each design phase, and sufficiently clear that the recipient for the next phase can easily understand the inputs to that task.

SIL 3 and SIL 4 require, also, that the project management identify the additional procedures and activities required at these levels and that there is a robust reporting mechanism to confirm both the completion and correctness of each activity. The documentation used for these higher SIL systems should be generated based on standards which give guidance on consistency and layout and include checklists. In addition, for SIL 4 systems, computer aided configuration control and computer aided design documentation should be used. Table B6 of the Standard elaborates a little on what constitutes a higher rigour of project management.

Much of the above 'good practice' (e.g. references to Project Management) tends to be repeated for each of the life-cycle activities, in both text and tables, throughout the Standard. We have attempted to avoid this repetition in this book. There are other aspects of the Standard's guidance which are repetitious and we have tended to refer to each item once in the most appropriate section.

The need for validation planning is stressed in Section 7.3 of the standard and this should be visible in the project Quality and Safety Plan which will include reference to the Functional Safety Audits (see also Section 9 of Appendix 1).

In general this whole section will be met by implementing the Functional Safety Procedure described in Appendix 1.

3.2 Requirements involving the specification

Section 7.2 of the Standard – Table B1 (avoidance)

(a) The specification for the system should be well structured, for all SILs, and cover:

- Integrity level requirement plus type of operation, i.e. low demand or high demand

- Safety function requirements
- System architecture
- Operational performance and modes of operation
- Interfaces with other systems and operators
- Environmental design requirements for the safety system equipment

Structured design techniques should be used at all SIL levels. At the system application level the functional requirements (i.e. logic) can be expressed by using semi-formal methods such as cause and effect diagrams or logic/function block diagrams. All this can be suitable up to SIL 3. For SIL 4 applications structured methods should be used. These include Yourdon, MASCOT, SADT, and several other techniques referenced in Part 7 of the Standard. In the case of new product design rather than applications engineering (i.e. design of executive software) structured methods should be progressively considered from SIL 2 upwards.

(b) Separation of functions. In order to reduce the probability of common cause failures the specification should also cover the degree of separation required, both physically and electrically, between the EUC and the safety system(s). Any necessary data interchange between the two systems should also be tightly specified.

These requirements need to be applied to any redundant elements of the safety-related system(s).

Achieving this separation may not always be possible since parts of the EUC may include a safety function that cannot be dissociated from the control of the equipment. This is more likely for the continuous mode of operation in which case the whole control system should be treated as safety-related pending target SIL calculations (Section 2.2).

If the safety-related and non-safety-related system elements cannot be shown to be sufficiently independent then the complete system should be treated as safety-related.

For SIL 1 and SIL 2 there should be a clear specification of the separation between the EUC and the safety system and electrical/data interfaces should be well defined. Physical separation should be considered.

For SIL 3 there should be physical separation between the EUC and the safety system and, also, the electrical/data interfaces should be clearly specified. Physical separation of redundant parts of the safety system should be considered.

For SIL 4 there should be total physical/electrical/data separation between the safety system and the EUC and between the redundant parts of the safety system.

3.3 Requirements for design and development

Section 7.4 of the Standard – Table B2 (avoidance)

3.3.1 Features of the design

Sections 7.4.1–7.4.9 excluding 7.4.3

(a) Use of in-house design standards and work practices needs to be evident. These will address proven components and parts, preferred designs and configurations etc.

(b) On manual or auto-detection of a failure the design should ensure system behaviour which maintains the overall safety targets. In general, this requires that failure in a safety system having redundant paths should be repaired within the mean time to repair assumed in the hardware reliability calculations. If this is not possible, then the procedure should be the same as for non-redundant paths as follows. On failure in the safety system with no redundant paths, either additional process monitoring should be provided to maintain adequate safety or the EUC should be shut down.

(c) Sector specific requirements need to be observed. Many of these are contained in the documents listed in Chapter 9.

(d) The system design should be structured and modular and use well-tried modules/components. Structured, in this context, implies clear partitioning of functions and a visible hierarchy of modules and their interconnection. For SIL 1 and SIL 2 the modularity should be kept to a 'limited size' and each module/component should have had previously documented field experience for at least one year with ten devices. If previous experience does not exist, or is insufficiently documented, then this can be replaced with additional modular/component testing. Such use of subjective descriptions (e.g. the 'limited size') adds further weight to the desirability of in-house checklists, which can be developed in the light of experience.

In addition for SIL 3 systems, previous experience is needed in a relevant application and for a period of at least two years with ten devices or, alternatively, some third party certification.

SIL 4 systems should be both proven in use, as mentioned above, and have third party certification.

It is worth mentioning that the 'years' of operation referred to above assume full time use (i.e. 8760 hrs per annum).

(e) Systematic failures caused by the design (this refers to Table A16 and A19 (control))

The primary technique is to use monitoring circuitry to check the functionality of the system. The degree of complexity required for this monitoring ranges from 'low' for SIL 1 and SIL 2, through 'medium' for SIL 3 to 'high' for SIL 4.

For example, a PLC-based safety system at either SIL 1 or SIL 2 would require, as a minimum, a watchdog function on the PLC CPU being the most complex element of this 'lower' integrity safety system.

These checks would be extended in order to meet SIL 3 and would include additional testing on the CPU (i.e. memory checks) along with basic checking of the I/O modules, sensors and actuators.

The coverage of these tests would need to be significantly increased for SIL 4 systems. Thus the degree of testing of input and output modules, sensors and actuators would be substantially increased. Again, however, these are subjective statements and standards such as IEC 61508 do not and cannot give totally prescriptive guidance. Nevertheless some guidance is given concerning diagnostic coverage.

> **It should be noted that the minimum configuration table given in Section 3.3.2 of this chapter permits higher SIL claims, despite lower levels of diagnosis, by virtue of either more redundancy or a higher proportion of 'fail safe' type failures.**

(f) Systematic failures caused by environmental stress (this refers to Table A17)

This requirement applies to all SIL levels and states that all components (indeed the overall system) should be designed and tested as suitable for the environment in question. This includes temperature and temperature cycling, EMC (electromagnetic compatibility), vibration, electrostatic etc. Components and systems that meet the appropriate IEC component standards, or CE marking, UL (Underwriters Laboratories Inc.) or FM (Factory Mutual) approval would generally be expected to meet this requirement.

(g) Systematic operation failures (this refers to Table A18 (control))
For all SILs the system should have protection against on-line modifications of either software or hardware.

There needs to be feedback on operator actions, particularly when these involve keyboards, in order to assist the operator in detecting mistakes.

As an example of this, for SIL 1 and SIL 2, all input operator actions should be repeated whereas, for SIL 3 and SIL 4, significant and consistent validation checks should be made on the operator action before acceptance of the commands.

The design should take into account human capabilities and limitations of operators and maintenance staff. Human factors are addressed in Chapter 6.4 of this book.

(h) Tables A1 to A16 of the Standard are techniques which the drafters (over a period of 10 years) considered suitable for achieving improvements in diagnostic capability. The following section (together with Appendix 4) discusses how to measure diagnostic capability and SFF. Should it then be necessary to enhance the diagnostic coverage, these tables can be used as a guide to techniques.

3.3.2 Architecture (i.e. safe failure fraction)

Section 7.4.3.1 – Tables '2' and '3'

Regardless of the hardware reliability calculated for the design, the standard specifies minimum levels of redundancy coupled with given levels of fault tolerance (described by the Safe Failure Fraction).

This safe failure fraction, for each safety function, needs to be estimated as shown in Appendix 4. The higher the SFF percentage requirement the more onerous is the demonstration.

For 60% the simple block architecture approach shown in Appendix 4 may be sufficient. For 90% and above, however, a more rigorous FMEA approach (also shown in the Appendix) is required.

The term SAFE FAILURE FRACTION (SFF) has been coined, in IEC 61508, to replace the earlier concept of diagnostic coverage. The percentages described as the 'safe failure fraction' refer to the sum of the potentially dangerous failures revealed by auto-test together with those which result in a safe state, as a fraction of the TOTAL number of failures. Thus:

$$\text{SFF} = \frac{\text{Total revealed hazardous failures} + \text{Total safe failures}}{\text{Total failures}}$$

('Total failures' are those on the top line PLUS the unrevealed hazardous failures.)

A 'fail safe' example might be a slamshut valve where 90% of the failures are 'spurious closure' and 10% 'fail to close'. In that case, a 90% 'safe failure fraction' would be claimed without further need to demonstrate automatic diagnosis. On the other hand, a combined example might be a control system whereby 50% of failures are 'fail-safe' and the remaining 50% enjoy an 80% automatic diagnosis. In this latter case the overall safe failure fraction becomes 90% (i.e. 50% + 0.8 × 50%).

There are two tables which cover the so-called 'Type A' components (failure modes well defined PLUS behaviour under fault conditions well defined PLUS failure data available) and the 'Type B' components (likely to be more complex and whereby any of the above are not satisfied).

In the following tables 'm' refers to the number of failures which lead to system failure. The tables provide the SIL number for each safe failure fraction case. The expression 'm + 1' implies redundancy whereby there are (m + 1) elements and m failures are sufficient to cause system failure.

TYPE A SFF	**SIL for Simplex**	**SIL for (m + 1)**	**SIL for (m + 2)**
<60%	1	2	3
60–90%	2	3	4
90–99%	3	4	4
>99%	3	4	4

TYPE B SFF	SIL for Simplex	SIL for $(m + 1)$	SIL for $(m + 2)$
<60%	NO*	1	2
60–90%	1	2	3
90–99%	2	3	4
>99%	3	4	4

- *This configuration is not allowed.
- Simplex infers no redundancy.
- $(m + 1)$ infers 1 out of 2, 2 out of 3 etc.
- $(m + 2)$ infers 1 out of 3, 2 out of 4 etc.

The above tables refer to 60%, 90% and 99%. At first this might seem a realistic range of safe fail fraction ranging from simple to comprehensive. However, it is worth considering how the diagnostic part of each of these coverage levels might be established. There are two ways in which diagnostic coverage and safe failure fraction ratios can be assessed:

1. **By test:** where failures are simulated and the number of diagnosed failures, or those leading to a safe condition, are counted.
2. **By FMEA:** where the circuit is examined to ascertain, for each potential component failure mode, whether it would be revealed by the diagnostic program or lead to a safe condition.

Clearly a 60% safe failure fraction could be demonstrated fairly easily by either method. Test would require a sample of only a few failures to reveal 60%, or alternatively a 'broad brush' FMEA, addressing blocks of circuitry rather than individual components, would establish (in an hour or two) if 60% were achieved. This is illustrated in Appendix 4.

Turning to 90% coverage, the test sample would now need to exceed 20 failures (for reasonable statistical significance) and the FMEA would require a more detailed approach. In both cases the cost and time become more significant. A fuller FMEA as illustrated in Appendix 4 is needed and might well involve 3–4 mandays.

For 99% coverage a reasonable sample size would now exceed 200 failures and the test demonstration is likely to be impracticable.

The foregoing should be considered carefully to ensure that there is adequate evidence to claim 90% and an even more careful examination before accepting the credibility of a 99% claim.

In order to take credit for diagnostic coverage, as described in the Standard (i.e. the above Architectural Constraint Tables), the time interval between repeated tests should at least be an order of magnitude less than the expected demand interval. For the case of a continuous system then the auto-test interval plus the time to put the system into a safe state should be within the time it takes for a failure to propagate to the hazard.

3.3.3 Random hardware failures

Section 7.4.3.2

This is better known as 'reliability prediction' which, in the past, has dominated risk assessment work. It involves specifying the reliability model, the failure rates to be assumed, the component down times, diagnostic intervals and coverage.

Techniques such as FMEA (failure mode and effect analysis) and fault tree analysis are involved and Chapters 6 and 7 briefly describe how to carry these out. The Standard refers to confidence levels in respect of failure rates and this will be dealt with later.

In Chapter 1 we mentioned the anomaly concerning the allocation of the quantitative failure probability target to the random hardware failures alone. There is yet another anomaly concerning judgement of whether the target is met. If the fully quantified approach (described in Chapter 2) has been adopted then the failure target will be a PFD (probability of failure on demand) or a failure rate. The reliability prediction might suggest that the target is not met although still remaining within the limits of the SIL in question. The rule here is that since we have chosen to adopt a fully quantitative approach we should meet the target set (paragraphs 7.2.3.2 and 7.4.3.2.1 of Part 2 of the Standard confirm this view). For example, a PFD of 2×10^{-3} might have been targeted for a safety-related risk reduction system. This is, of course, SIL 2. The assessment might suggest that it will achieve 5×10^{-3} which is indeed SIL 2. However,

we chose to set a target of 2×10^{-3} and therefore have NOT met it.

The question might then be asked 'What if we had opted for a simpler risk graph approach and stated the requirement merely as a SIL – then would we not have met the requirement?' This appears to be inconsistent. Once again there is no right or wrong answer to the dilemma. The Standard does not address it and, as in all such matters, the judgement of the responsible engineer is needed. Both approaches are admissible and, in any case, the accuracy of quantification is not very high (see Chapter 7).

3.4 Integration and test (referred to as verification)

Sections 7.5 and 7.9 of the Standard – Table B3 (avoidance)

Based on the intended functionality the system should be tested, and the results recorded, to ensure that it fully meets the requirements. This is the type of testing which, for example, looks at the output responses to various combinations of inputs. This applies to all SILs.

Furthermore, a degree of additional testing, such as the response to unusual and 'not specified' input conditions, should be carried out. For SIL 1 and SIL 2 this should include system partitioning testing and boundary value testing. For SIL 3 and SIL 4 the tests should be extended to include test cases that combine critical logic requirements at operation boundaries.

3.5 Operations and maintenance

Section 7.6 – Table B4 (avoidance)

(a) The system should have clear and concise operating and maintenance procedures. These procedures, and the safety system interface with personnel, should be designed to be user, and maintenance, friendly. This applies to all SIL levels.

(b) Documentation needs to be kept, of audits and for any proof-testing that is called for. There need to be records of the demand rate of the safety-related equipment, and furthermore failures also need to be recorded. These records should be periodically reviewed, to verify that the target safety integrity level was indeed appropriate and that it has been achieved. This applies to all SILs.

(c) For SIL 1 and SIL 2 systems, the operator input commands should be protected by key switches/passwords and all personnel should receive basic training. In addition, for SIL 3 and SIL 4 systems operating/maintenance procedures should be highly robust and personnel should have a high degree of experience and undertake annual training. This should include a study of the relationship between the safety-related system and the EUC.

3.6 Validation (meaning overall acceptance test and close-out of actions)

Section 7.7 – Table B5 (avoidance)

The object is to ensure that all the requirements of the safety system have been met and that all the procedures have been followed (albeit this should be ensured as a result of functional safety capability). It is necessary to ensure that any remedial action or additional testing arising from earlier tests has been carried out. This requirement applies to all SIL levels.

3.7 Modifications

Section 7.8

For all modifications and changes there should be:

- revision control
- a record of the reason for the design change
- an impact analysis
- retesting of the changed and any other affected modules

The methods and procedures should be exactly the same as those applied at the original design phase. This paragraph applies to all SILs.

Part 3 of the Standard (Chapter 4 of this book) requires that for SIL 1 changed modules are reverified, for SIL 2 all affected modules are reverified and for SIL 3 the whole system is revalidated. Although Part 2 does not specify this for the hardware the authors consider this to be good practice.

3.8 Acquired sub-systems

Any sub-system which is to be used as part of the safety system, and is acquired as a complete item by the integrator of the safety system, will need to establish, in addition to any other engineering considerations, the following parameters.

- Random hardware failure rates, categorised as:
  - fail safe failures
  - dangerous failures detected by auto-test
  - dangerous failures detected by proof test
- Procedures/methods for adequate proof testing
- The hardware fault tolerance of the sub-system
- The highest SIL that can be claimed as a consequence of the measures and procedures used during the design and implementation of the hardware and software, or
- A SIL derived by claim of 'proven in use' see Section 3.9 below

3.9 'Proven in use'

As an alternative to all the systematic requirements summarised in this chapter, adequate statistical data from field use may be used to satisfy the Standard. The random hardware failures prediction and safe failure fraction demonstrations are, however, still required. The previous field experience should be in an application and environment, which is very similar to the intended use. All failures experienced, whether due to hardware failures or systematic faults, should be recorded, along with total running hours. The Standard asks that the calculated failure rates should be claimed using a confidence limit of at least 70%.

Paragraphs 7.4.7.5 to 7.4.7.12 of Part 2 allow for statistical demonstration that a SIL has been met in use. In Part 7 Annex D there are a number of pieces of statistical theory which purport to be appropriate to establishing confidence for software failures. However, the same theory applies to hardware failures and for the purposes of the single-sided 70% requirement can be summarised as follows.

For zero failures, the following 'number of operations/demands' or 'equipment hours' are necessary to infer that the lower limit of each SIL has been exceeded:

SIL 1 (1: 10^{-1} or 10^{-1} per annum) 12 operations or 12 years
SIL 2 (1: 10^{-2} or 10^{-2} per annum) 120 operations or 120 years
SIL 3 (1: 10^{-3} or 10^{-3} per annum) 1200 operations or 1200 years
SIL 4 (1: 10^{-4} or 10^{-4} per annum) 12 000 operations or 12 000 years

For one failure, the following table applies. The times for larger numbers of failures can be calculated accordingly (i.e. from chi square methods).

SIL 1 (1: 10^{-1} or 10^{-1} per annum) 24 operations or 24 years
SIL 2 (1: 10^{-2} or 10^{-2} per annum) 240 operations or 240 years
SIL 3 (1: 10^{-3} or 10^{-3} per annum) 2400 operations or 2400 years
SIL 4 (1: 10^{-4} or 10^{-4} per annum) 24 000 operations or 24 000 years

3.10 Presenting the results

In order to justify that the SIL requirements have been correctly selected and satisfied, it is necessary to provide a documented assessment.

The following Conformance Demonstration Template is suggested as a possible format based on the layout of this chapter.

The Standard, in Part 6, gives two examples of Part 3 assessments. It does not, however, provide these for the Part 2 requirements.

Conformance Demonstration Template
IEC 61508 Part 2

For embedded software designs, with new hardware design, the demonstration might involve a reprint of all the tables from the Standard. The evidence for each item would then be entered in the right-hand column as in the simple tables below.

The following tables might be considered adequate for relatively simple designs, particularly with existing platforms and simple low variability code as in the case of PLCs.

Under '**Evidence**' enter a reference to the project document (e.g. spec, test report, review, calculation) which satisfies that requirement.

Under '**Feature**' take the text in conjunction with the fuller text in this chapter.

General

(Paras 7.1 & 7.3) **(Table '1')**

Feature (all SILs)	**Evidence**
Existence of Quality and Safety Plan, including document hierarchy, tasks and competency etc. Description of overall novelty, complexity, SILs, rigour needed etc. Clear documentation hierarchy (Q & S Plan, Functional Spec, Design docs, Review strategy, Integration and test plans etc.) Adequate project management as per company's FSCA procedure	
Feature (SIL 3 and above)	**Evidence**
Enhanced rigour of project management	

Life-cycle

(Paras 7.1 & 7.3) **(Table '1')**

Feature (all SILs)	**Evidence**
A Functional Safety audit has given a reasonable indication that the life-cycle activities required by the company's FSCA procedure have been implemented The project plan should include adequate plans to validate the overall requirements and state tools and techniques	

Specification

(Para. 7.2) **(Table B1)**

Feature (all SILs)	**Evidence**
Clear text and some graphics, use of checklist or structured method, precise, unambiguous. Describes SR functions and separation of EUC/SRS, responses, performance requirements, well-defined interfaces, modes of operation SIL for each SR function, high/low demand, proof test, emc	
Feature (SIL 2 and above)	**Evidence**
Inspection of the specification Either computer aided spec tool or structured method	

Feature (SIL 3 and above)	Evidence
Use of a semi-formal method Physical separation of EUC/SRS	
Feature (SIL 4)	**Evidence**
Independent validation of the spec, total 'write' separation of SR and other functions, separate location SR and other functions, computer aided documentation Physical separation of redundant elements	

Design and development

(Para. 7.4 (excl. 7.4.3)) **(Tables B2, A16–A18)**

Feature (all SILs)	Evidence
Use of in-house design standards and work instructions Sector specific guidance as required Visible and adequate design documentation Structured design Proven components and sub-systems (justified by 10 for 1 year) Modular approach with SR elements independent of non-SR and interfaces well defined SR SIL = Highest of mode SILs Adequate component derating Non-SR failures independent of SRS Safe state achieved on detection of failure Data communications errors addressed No access by user to change hardware or software Operator interfaces considered Fault tolerant technique (minimum of a watchdog) Appropriate emc measures	
Feature (SIL 2 and above)	**Evidence**
Checklist or walkthrough or design tools Higher degree of fault tolerance Appropriate emc measures as per Table A17	
Feature (SIL 3 and above)	**Evidence**
Use of semi-formal methods Proven components and sub-systems (certified or justified by 10 for 2 years) Higher degree of fault tolerance and monitoring (e.g. memory checks)	

Feature (SIL 4)	Evidence
Higher degree of fault tolerance and monitoring Use of additional checking tools or methods Proven components and sub-systems (certified AND justified by 10 for 2 years)	

Random hardware failures and architectures

(Para. 7.4.3.1 & 2)

Feature (all SILs)	Evidence
SFF and architectural conformance is to be demonstrated Random hardware failures are to be predicted and compared with the SIL or other quantified target Random hardware failures assessment contains all the items suggested in Appendix 2 of this book	
Feature (SFF $\geqslant$90%)	**Evidence**
SFF assessed by a documented FMEA Appropriate choice of A or B type table	

Integration and test

(Paras 7.5 & 7.9) **(Table B3)**

Feature (all SILs)	Evidence
Overall test strategy in Q & S plan Test specs, logs of results and discrepancies, records of versions, acceptance criteria, tools and remedial action Functional test including input partitioning, boundary values and non-specified input values	
Feature (SIL 2 and above)	**Evidence**
As for SIL 1	
Feature (SIL 3 and above)	**Evidence**
Include tests of critical logic functions at operational boundaries Standardised procedures	
Feature (SIL 4)	**Evidence**
Distribution of input data according to real life application	

Operations and maintenance

(Para. 7.6) (Table B4)

Feature (all SILs)	Evidence
Component wear-out life accounted for by preventive replacement Proof tests specified Procedures validated by Ops/Mtce staff Commissioning successful Reporting procedures in place User friendly interfaces Lockable switch or password access Operator i/ps to be acknowledged Basic training	
Feature (SIL 2 and above)	**Evidence**
As SIL 1	
Feature (SIL 3 and above)	**Evidence**
More robust procedures At least annual training	
Feature (SIL 4)	**Evidence**
Consistency checks on every command Five years' experience for operators	

Validation

(Para 7.7) (Table B5)

Feature (all SILs)	Evidence
Validation plan actually implemented Calibration of equipment Records and close-out report Discrepancies positively handled Functional tests Environmental tests Interference tests Fault insertion when diagnostic target >90%	
Feature (SIL 2 and above)	**Evidence**
Check all SR functions OK in presence of faulty operating conditions	

Feature (SIL 3 and above)	Evidence
Fault insertion at unit level Some static or dynamic analysis or simulation	
Feature (SIL 4)	**Evidence**
Use of static/dynamic analysis tools Rare worst case and boundary value tests Fault insertion at component level	

Modifications

(Para. 7.8)

Feature (all SILs)	Evidence
Change control with adequate competence Impact analysis Reverify changed modules	
Feature (SIL 2 and above)	**Evidence**
Reverify affected modules	
Feature (SIL 3 and above)	**Evidence**
Revalidate whole system	

Acquired sub-systems

Feature (at the appropriate SIL)	Evidence
SIL requirements reflected onto suppliers Compliance demonstrated	

Proven in use

(Paras 7.4.7.5–12)

Feature (at the appropriate SIL)	Evidence
Application appropriate and restricted functionality Any differences to application addressed and conformance demonstrated Statistical data available at 70% confidence to verify random hardware failures target Failure data validated	

CHAPTER 4

MEETING IEC 61508 PART 3

IEC 61508 Part 3 covers the development of software. Sections 4.1–4.8 summarise the main requirements. However, the following points should be noted first.

Whereas the reliability prediction of hardware failures, as addressed in Section 3.3.3 of the last chapter, implies a failure rate to be anticipated, the application of qualitative measures DOES NOT infer a failure rate for the systematic failures. All that can be reasonably claimed is that, given the state of the art, we believe the measures specified are appropriate for the integrity level in question and that therefore the systematic failures will probably be similar to and not exceed the hardware failure rate of that SIL.

The Annexes of Part 3 offer appropriate techniques, by SIL, in the form of tables followed by more detailed tables with cross-references.

This chapter attempts to provide a simple and usable interpretation. At the end of this chapter a 'conformance demonstration template' is suggested which, when completed for a specific product or system assessment, will provide evidence of conformance to the SIL in question.

> The approach to the assessment will differ substantially between:
>
> EMBEDDED SOFTWARE DESIGN (CASS Type 1 (see Chapter 10))
>
> and
>
> APPLICATIONS SOFTWARE (CASS Type 2 (see Chapter 10))
>
> The demonstration template tables at the end of this chapter cater for the latter case (i.e. Type 2). Chapter 5, using the restricted subset of IEC 61511, also caters for the Type 2 case.

4.1 Organising and managing the software engineering

Sections 7.1 and 7.3 of the Standard – Table '1'

Section 3.1 of the previous chapter applies here in exactly the same way and therefore we do not repeat it.

In addition, the Standard recommends the use of the 'V' model approach to software design, with the number of phases in the 'V' model being adapted according to the target safety integrity level and the complexity of the project. The principle of the 'V' model is a top-down design approach starting with the 'overall software safety specification' and ending, at the bottom, with the actual software code. Progressive testing of the system starts with the lowest level of software module, followed by integrating modules, and working up to testing the complete safety system. Normally, a level of testing for each level of design would be required.

The life-cycle should be described in writing (as well as graphical figures such as are shown in Figures 4.1–4.3). System and hardware interfaces should be addressed and it should reflect the architectural design.

At SIL 2 and above there needs to be evidence of positive justifications and reviews of departures from the life-cycle activities listed in the Standard.

Figures 4.2 and 4.3 show typical interpretations of this model as they might apply to the two types of development mentioned in the box at the beginning of this chapter. Beneath each of the figures is a statement describing how they meet the activities specified in the Standard.

Figure 4.2 describes a simple proven PLC platform with ladder logic code providing an application such as process control or shutdown. Figure 4.3 describes a more complex development where the software has been developed in a high level language (for example, C or Ada) and where there is an element of assembler code.

The software configuration management process needs to be clear and to specify:

- Levels where configuration control commences
- Where baselines will be defined and how they will be established

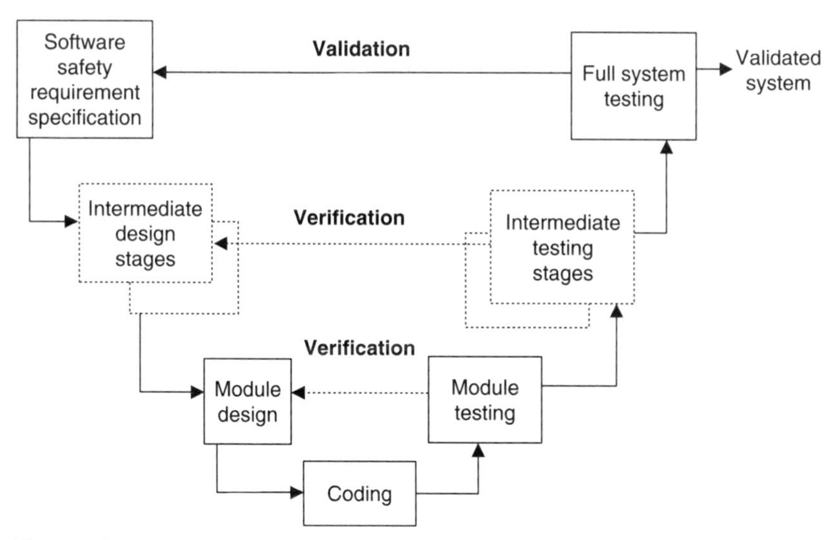

Figure 4.1
A typical 'V' model

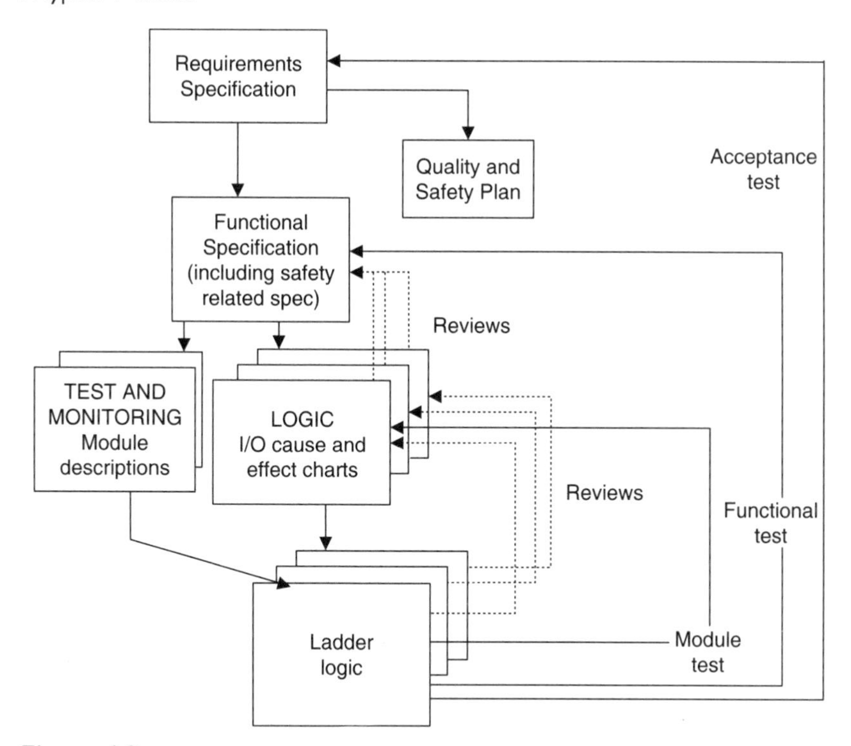

Figure 4.2
A software development life-cycle for a simple PLC system at the application level

- Methods of traceability of requirements
- Change control
- Impact assessment
- Rules for release and disposal

At SIL 2 and above configuration control must apply to the smallest compiled module or unit.

The life-cycle model in Figure 4.2 addresses the architectural design in the Functional Specification and the module design by virtue of cause and effect charts. Integration is a part of the functional test and validation is achieved by means of acceptance test and other activities listed in the Quality and Safety Plan.

The life-cycle model in Figure 4.3 addresses the architectural design in the Functional Specification. Validation is achieved by

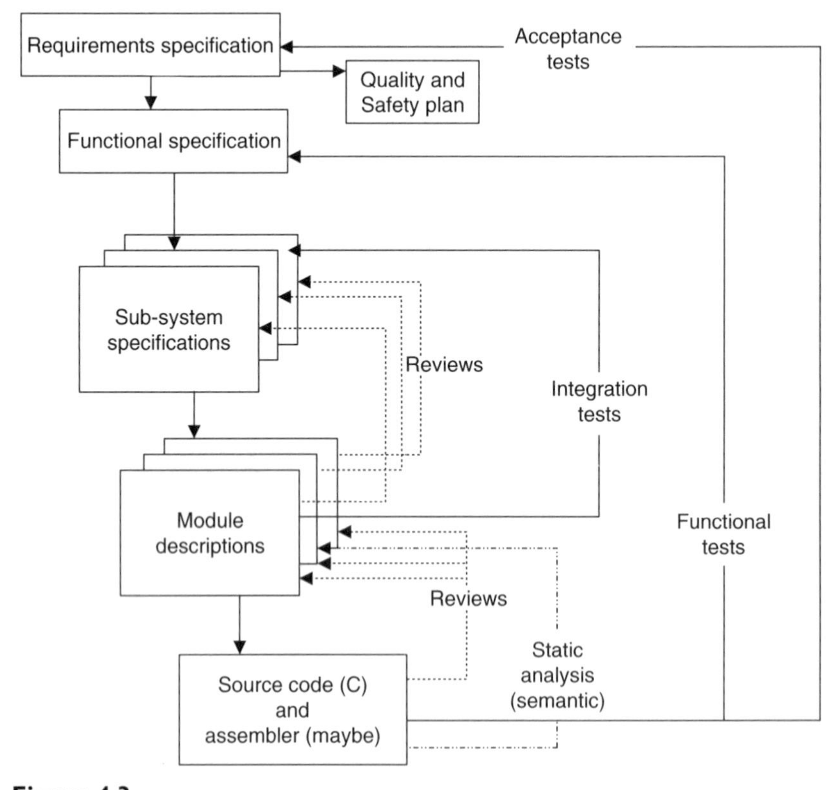

Figure 4.3
A software development life-cycle for a system with embedded software

means of acceptance test and other activities listed in the Quality and Safety Plan.

4.2 Requirements involving the specification

Section 7.2 of the Standard – Table A1

(a) The software safety requirements, in terms of both the safety functions and the safety integrity, should be stated in the software safety requirements specification.

(b) The specification should **include** all the modes of operation, the capacity and response time performance requirements, maintenance and operator requirements, self monitoring of the software and hardware as appropriate, enabling the safety function to be testable whilst the EUC is operational, and details of all internal/external interfaces. The specification should extend down to the configuration control level.

(c) The specification should be written in **a clear and precise** manner, traceable back to the safety specification and other relevant documents. The document should be free from ambiguity and clear to those for whom it is intended.

For SIL 1 and SIL 2 systems, this specification should use semi-formal methods to describe the critical parts of the requirement (e.g. safety-related control logic). For SIL 3 and SIL 4, semi-formal methods should be used for all the requirements and, in addition, at SIL 4 there should be the use of computer support tools for the critical parts (e.g. safety-related control logic).

The semi-formal methods chosen should be appropriate to the application and typically include logic/function block diagrams, cause and effect charts, sequence diagrams, state transition diagrams, time Petri nets, truth tables and data flow diagrams.

4.3 Requirements for design and development

4.3.1 Features of the design

Section 7.4.3.2 of the Standard – Table A2

(a) The design methods should aid modularity and embrace features which reduce complexity and provide clear expression of functionality, information flow, data structures, sequencing, timing-related constraints/information, and design assumptions.

(b) The system software (i.e. non-application software) should include software for diagnosing faults in the system hardware, error detection for communication links, and on-line testing of standard application software modules.

In the event of detecting an error or fault the system should, if appropriate, be allowed to continue but with the faulty redundant element or complete part of the system isolated.

For SIL 1 and SIL 2 systems there should be basic hardware fault checks (i.e. watchdog and serial communication error detection).

For SIL 3 and SIL 4, there needs to be some hardware fault detection on all parts of the system, i.e. sensors, input/output circuits, logic resolver, output elements and both the communication and memory should have error detection.

4.3.2 Detailed design

Paragraphs 7.4.5, 7.4.6 – Table A4

(a) The detailed design of the software modules and coding implementation should result in small manageable software modules. Semi-formal methods should be applied, together with design and coding standards including structured programming, suitable for the application. This applies to all SILs.

(b) The system should, as far as possible, use trusted and verified software modules, which have been used in similar applications. This is called for from SIL 2 upwards.

(c) The software should not use dynamic objects, which depend on the state of the system at the moment of allocation, where they do not allow for checking by off-line tools. This applies to all SILs.

(d) For SIL 3 and SIL 4 systems, the software should include additional defensive programming (e.g. variables should be both range and, where possible, plausibility checked). There should also be limited use of interrupts, pointers, and recursion.

4.3.3 Programming language and support tools

Paragraph 7.4.4 – Table A3

(a) The programming language should be capable of being fully and unambiguously defined. The language should be used with a specific coding standard and a restricted sub-set, to minimise unsafe/unstructured use of the language. This applies to all SILs.

At SIL 2 and above, dynamic objects and unconditional branches should be forbidden. At SIL 3 and SIL 4 more rigorous rules should be considered such as the limiting of interrupts and pointers, and the use of diverse functions to protect against errors which might arise from tools.

(b) The support tools need to be either well proven in use (and errors resolved) and/or certified as suitable for safety system application. The above applies to all SILs, with certified tools more strongly recommended for SIL 3 and SIL 4.

4.4 Integration and test (referred to as verification)

4.4.1 Software module testing and integration

Paragraphs 7.4.7, 7.4.8 – Table A5

(a) The individual software modules should be code reviewed and tested to ensure that they perform the intended function and by a selection of limited test data to confirm that the system does not perform unintended functions.

(b) As the module testing is completed then module integration testing should be performed with pre-defined test cases and test data. This testing should include functional, 'black box' and performance testing.

(c) The results of the testing should be documented in a chronological log and any necessary corrective action specified. Version numbers of modules and of test instructions should be clearly indicated. Discrepancies from the anticipated results should be clearly visible. Any modifications or changes to the software, which are implemented after any phase of the testing, should be analysed to determine the full extent of retest that is required.

(d) The above needs to be carried out for all SILs; however, the extent of the testing for unexpected and fault conditions needs to be increased for the higher SILs. As an example, for SIL 1 and SIL 2 systems the testing should include boundary value testing and partitioning testing and in addition, for SIL 3 and SIL 4, tests generated from cause consequence analysis of certain critical events.

4.4.2 Overall integration testing

Paragraph 7.5 – Table A6

These recommendations are for testing the integrated system, which includes both hardware and software, and although this requirement is repeated in Part 3 the same requirements have already been dealt with in Part 2.

4.5 Validation (meaning overall acceptance test and close-out of actions)

Paragraphs 7.7, 7.9 – Tables A7, 9

(a) Whereas verification implies confirming, for each stage of the design, that all the requirements have been met prior to the start of testing of the next stage (shown in Figures 4.1–4.3), validation is the final confirmation that the total system meets all the required objectives and that all the design procedures have been followed. The Functional Safety Management requirements (Chapter 2) should cover the requirements for both validation and verification.

(b) The validation plan should show how all the safety requirements have been fully addressed. It should cover the entire lifecycle activities and will show audit points. It should address specific pass/fail criteria, a positive choice of validation methods and a clear handling of non-conformances.

(c) At SIL 2 and above some test coverage metric should be visible. At **SILs 3 and 4** a more rigorous coverage of accuracy, consistency, conformance with standards (e.g. coding rules) is needed.

4.6 Modifications

Paragraph 7.8 – Table A8

(a) The following are required:

- A modification log
- Revision control
- Record of the reason for design change
- Impact analysis
- Retesting as in (b) below

The methods and procedures should be at least equal to those applied at the original design phase. This paragraph applies for all SIL levels.

The modification records should make it clear which documents have been changed and the nature of the change.

(b) For SIL 1 changed modules are reverified, for **SIL 2** all affected modules are reverified and for **SIL 3 and above** the whole system is revalidated.

4.7 Some technical comments

4.7.1 Static analysis

Static analysis is a technique (usually automated) which does not involve execution of code but consists of algebraic examination of source code. It involves a succession of 'procedures' whereby the paths through the code, the use of variables and the algebraic functions of the algorithms are analysed. There are packages available which carry out the procedures and, indeed, modern compilers frequently carry out some of the static analysis procedures such as data flow analysis.

Table B8 of Part 3 lists Data flow and Control flow as HR (highly recommended) for SIL 3 and SIL 4. It should be remembered, however, that static analysis packages are only available for procedural high-level languages and require a translator that is language specific. Thus, static analysis cannot be automatically applied to PLC code other than by means of manual code walkthrough which loses the advantages of the 100% algebraic capability of an automated package.

Semantic analysis, whereby functional relationships between inputs and outputs are described for each path, is the most powerful of the static analysis procedures. It is, however, not trivial and might well involve several mandays of analysis effort for a 500 line segment of code. It is not referred to in the Standard.

Static analysis, although powerful, is not a panacea for code quality. It only reflects the functionality in order for the analyst to review the code against the specification. Furthermore it is concerned only with logic and cannot address timing features.

It is worth noting that, in Table B8, design review is treated as an element of static analysis.

4.7.2 Use of 'formal' methods

Table B5 of Part 3 refers to formal methods and Table A9 to formal proof. In both cases it is HR (highly recommended) for SIL 4 and merely R (recommended) for SIL 2 and SIL 3.

The term formal methods is much used and much abused. In software engineering it covers a number of methodologies and techniques for specifying and designing systems, both non-programmable and programmable. These can be applied throughout the life-cycle including the specification stage and the software coding itself.

The term is often used to describe a range of mathematical notations and techniques applied to the rigorous definition of system requirements which can then be propagated into the subsequent design stages. The strength of formal methods is that they address the requirements at the beginning of the design cycle. One of the main benefits of this is that formalism applied at this early stage may lead to the prevention, or at least early detection, of incipient errors. The cost of errors revealed at this stage is dramatically less than if they are allowed to persist until commissioning or even field use. This is because the longer they remain undetected the potentially more serious and far reaching are the changes required to correct them.

The potential benefits may be considerable but they cannot be realised without properly trained people and appropriate tools. Formal methods are not easy to use. As with all languages, it is easier to read a piece of specification than it is to write it. A further complication is the choice of method for a particular application. Unfortunately, there is not a universally suitable method for all situations.

4.7.3 PLCs (Programmable Logic Controllers) and their languages

In the past, PLC programming languages were limited to simple code (e.g. Ladder Logic) which is a limited variability language usually having no branching statements. These earlier languages are suitable for use at all SILs with only minor restrictions on the instruction set.

Currently PLCs have wider instruction sets, involving branching instructions etc., and restrictions in the use of the language set are needed at the higher SILs.

With the advent of IEC 61131-3 there is a range of limited variability programming languages and the choice will be governed partly by the application. Again restricted subsets may be needed for safety-related applications. Some application specific languages are now available, as, for example, the facility to program plant shutdown systems directly by means of Cause and Effect Diagrams. Inherently, this is a restricted sub-set created for safety-related applications.

The IEE SEMSPLC Guidelines (described in Chapter 9) provide some more detail, although they date from 1996 and have now been withdrawn.

4.7.4 Software reuse

Parts 2 and 3 of the Standard refer to 'trusted/verified', 'proven in use' and 'field experience' in various tables and in parts of the text. They are used in slightly different contexts but basically refer to the same concept of empirical evidence from use. However, 'trusted/verified' also refers to previously designed and tested software without regard for its previous application and use.

Table A4 of Part 3 lists the reuse of 'trusted/verified' software modules as 'highly recommended' for SIL 2 and above.

It is frequently assumed that the reuse of software, including specifications, algorithms and code will, since the item is proven, lead to fewer failures than if the software were developed anew. There are reasons for and against this assumption.

Reasonable expectations of reliability, from reuse, are suggested because:

- The reused code or specification is proven.
- The item has been subject to more than average test.
- The time saving can be used for more development or test.
- The item has been tested in real applications environments.
- If the item has been designed for reuse it will be more likely to have stand-alone features such as less coupling.

On the other hand:

- If the reused item is being used in a different environment undiscovered faults may be revealed.
- If the item has been designed for reuse it may contain facilities not required for a particular application therefore the item may not be ideal for the application and it may have to be modified.
- Problems may arise from the internal operation of the item not being fully understood.

In conclusion, provided that there is adequate control involving procedures to minimise the effects of the above then significant advantages can be gained by the reuse of software at all SILs.

4.7.5 Software metrics

The term metrics, in this context, refers to measures of size, complexity and structure of code. An obvious example would be the number of branching statements (in other words a measure of complexity) which might be assumed to relate to error rate. There has been interest in this activity for many years but there are conflicting opinions as to its value.

Table A9 of Part 3 mentions software metrics but merely lists them as 'recommended' at all SILs. In the long term metrics, if collected extensively within a specific industry group or product application, might permit some correlation with field failure performance and safety-integrity. It is felt, however, that it is still 'early days' in this respect.

The term metrics is also used to refer to statistics about test coverage, as called for in earlier paragraphs.

4.8 'Proven in use'

Software that is to be reused needs to meet the entire specification requirement for its intended function. It should also have satisfied suitable procedures, testing and verification for the SIL in question or have evidence to support its use from satisfactory previous use. 'Proven in use' needs to show that the previous use is to the same specification with an operating experience of at least three to five times greater than the period between demands. Also, the operating experience should have exercised all the safety-related functions associated with the module. These claims need to have been documented during the period claimed for 'proven in use' and there should have been no safety-related failures.

In Part 3, Paragraphs 7.4.2.11 and 7.4.7.2 (Note 3) allow for statistical demonstration that an SIL has been met in use for a module of software. In Part 7 Annex D there are a number of pieces of statistical theory which purport to be appropriate to the confidence in software. However, the same theory applies as with hardware failure data and for the purposes of the single-sided 70% requirement can be summarised as follows.

For zero failures, the following 'number of operations/demands' or 'equipment hours' are necessary to infer that the lower limit of each SIL has been exceeded. This is the same as was given for hardware, in Chapter 3:

SIL 1 (1: 10^{-1} or 10^{-1} per annum) 12 operations or 12 years
SIL 2 (1: 10^{-2} or 10^{-2} per annum) 120 operations or 120 years
SIL 3 (1: 10^{-3} or 10^{-3} per annum) 1200 operations or 1200 years
SIL 4 (1: 10^{-4} or 10^{-4} per annum) 12 000 operations or 12 000 years

For one failure, the following table applies. The times for larger numbers of failures can be calculated accordingly (i.e. from chi square methods):

SIL 1 (1: 10^{-1} or 10^{-1} per annum) 24 operations or 24 years
SIL 2 (1: 10^{-2} or 10^{-2} per annum) 240 operations or 240 years
SIL 3 (1: 10^{-3} or 10^{-3} per annum) 2400 operations or 2400 years
SIL 4 (1: 10^{-4} or 10^{-4} per annum) 24 000 operations or 24 000 years

4.9 Presenting the results

In order to justify that the SIL requirements have been correctly selected and satisfied, it is necessary to provide a documented assessment.

The following Conformance Demonstration Template is suggested as a possible format based on the layout of this chapter.

The Standard, in Part 6, gives two examples of Part 3 assessments.

Conformance Demonstration Template
IEC 61508 Part 3

For embedded software designs, with new hardware design, the demonstration might involve a reprint of all the tables from the Standard. The evidence for each item would then be entered in the right-hand column as in the simple tables below.

The following tables might be considered adequate for relatively simple designs, particularly with existing platforms and simple low variability code as in the case of PLCs.

Under '**Evidence**' enter a reference to the project document (e.g. spec, test report, review, calculation) which satisfies that requirement.

Under '**Feature**' take the text in conjunction with the fuller text in this chapter.

General

(Paras 7.1 & 7.3) **(Table '1')**

Feature (all SILs)	**Evidence**
Existence of Quality and Safety Plan, including document hierarchy, tasks and competency etc.	
Description of overall novelty, complexity, SILs, rigour needed etc.	
Clear documentation hierarchy (Q&S Plan, Functional Spec, Design docs, Review strategy, Integration and test plans etc.)	
Adequate project management as per company's FSCA procedure	
Feature (SIL 3 and above)	**Evidence**
Enhanced rigour of project management	

Life-cycle

(Paras 7.1 & 7.3) **(Table '1')**

Feature (all SILs)	Evidence
A Functional Safety audit has given a reasonable indication that the life-cycle activities required by the company's FSCA procedure have been implemented The project plan should include adequate plans to validate the overall requirements and state tools and techniques Adequate software life-cycle model as per Section 4.1 of this chapter Configuration management specifying baselines, minimum configuration stage, traceablity, release etc.	
Feature (SIL 2 and above)	**Evidence**
Alternative life-cycle models to be justified Configuration control to level of smallest compiled unit	
Feature (SIL 3 and above)	**Evidence**
Alternative life-cycle models to be justified and at least as rigorous Sample review of configuration status	
Feature SIL 4	**Evidence**
Alternative measures to the life-cycle to be separately reviewed	

Specification

(Para. 7.2) **(Table A1)**
(Table B7 amplifies semi-formal methods)

Feature (all SILs)	Evidence
Clear text and some graphics, use of checklist or structured method, precise, unambiguous and traceable. Describes SR functions and their separation, performance requirements, well-defined interfaces, modes of operation Capacities and response times declared SIL for each SR function, high/low demand, proof test, emc Self monitoring and self test features A review of the feasibility of requirements	

Feature (SIL 2 and above)	Evidence
Inspection of the specification (traceability to interface specs) Either computer aided spec tool or semi-formal method	
Feature (SIL 3 and above)	**Evidence**
Use of a semi-formal method (i.e. systematic representation of the logic throughout the spec)	
Feature (SIL 4)	**Evidence**
Use of a formal method	

Architecture and fault tolerance

(Para. 7.4.3) **(Table A2)**

Feature (all SILs)	Evidence
Major elements of the software, and their interconnection, defined Clear partitioning into functions Use of structured methods in describing the architecture Address graceful degradation (i.e. resilience to faults) Program sequence monitoring (i.e. a watchdog function)	
Feature (SIL 2 and above)	**Evidence**
Clear visibility of logic (i.e. the algorithms)	
Feature (SIL 3 and above)	**Evidence**
Fault detection and diagnosis Program sequence monitoring (i.e. counters and memory checks) Use of a semi-formal method	
Feature (SIL 4)	**Evidence**
Use of a formal method Increased error detection AND correction Diverse programming (for redundancy)	

Design and development

(Paras 7.4.5, 7.4.6) **(Tables A4, B1, B9)**

Feature (all SILs)	Evidence
Use of standards and guidelines Visible and adequate design documentation	

Modular design with minimum complexity Small manageable modules (and modules conform to the coding standards) Diagnostic software (e.g. watchdog and comms checks) Isolate and continue on detection of fault Structured methods	
Feature (SIL 2 and above) Trusted and verified modules No dynamic objects No unconditional jumps	**Evidence**
Feature (SIL 3 and above) Additional defences (e.g. range checks) No (or on-line check) dynamic variables Limited pointers, interrupts, recursion	**Evidence**
Feature (SIL 4) A formal method Protection against corruption from non-SIL4 equipment Semantic review of functionality	**Evidence**

Language and support

(Para. 7.5) **(Table A3)**

Feature (all SILs) Suitable language Language fully defined Coding standard/manual (fit for purpose) Confidence in tools	**Evidence**
Feature (SIL 2 and above) Trusted module library No dynamic objects	**Evidence**
Feature (SIL 3 and above) Certified tools or proven in use to be error free Language sub-set (e.g. limited interrupts and pointers)	**Evidence**
Feature (SIL 4) Certified AND proven tools Full assessment of calculation precision, execution order, exception handling	**Evidence**

Integration and test

(Paras 7.4.7, 7.4.8 & 7.5) (Tables A5, A6, B2, B3)

Feature (all SILs)	Evidence
Overall test strategy in Q&S Plan including provision for remedial action	
Test specs, results and discrepancy records and remedial action evidence	
Test logs in chronological order with version referencing	
Module code review	
Pre-defined test cases with boundary values	
Response times and memory constraints	
Functional and black box testing	
Feature (SIL 2 and above)	**Evidence**
Dynamic analysis	
Unintended functions tested on critical paths	
Feature (SIL 3 and above)	**Evidence**
Tests based on cause consequence analysis	
Avalanche/stress tests	
Feature (SIL 4)	**Evidence**
Probabilistic testing (statistical analysis of test coverage and results)	
Input partitioning testing	

Validation

(Paras 7.7, 7.9) (Tables A7, A9, B5, B8)

Feature (all SILs)	Evidence
Validation plan exists and is actually implemented	
Calibration of equipment	
Records and close-out report	
Suitable and justified choice of methods and models	
Feature (SIL 2 and above)	**Evidence**
Static analysis	
Test case metrics	
Feature (SIL 3 and above)	**Evidence**
Simulation or modelling	
Further reviews (e.g. dead code, test coverage adequacy, behaviour of algorithms)	

Feature (SIL 4)	Evidence
Probabilistic testing (statistical analysis of test coverage) A process of formal proof	

Modifications

(Para. 7.8) (Table B8)

Feature (all SILs)	Evidence
Modification log Change control with adequate competence Software configuration management Impact analysis Reverify changed modules	
Feature (SIL 2 and above)	**Evidence**
Reverify affected modules	
Feature (SIL 3 and above)	**Evidence**
Revalidate whole system	

Acquired sub-systems

Feature (at the appropriate SIL)	Evidence
SIL requirements reflected onto suppliers	

Proven in use

(Paras 7.4.2, 7.4.7)

Feature (at the appropriate SIL)	Evidence
Application appropriate Statistical data available Failure data validated	

Assessment

(Para. 8) (Tables A10, B4)

Feature (all SILs)	Evidence
Either checklists, truth tables, or block diagrams	
Feature (SIL 2 and above)	**Evidence**
As SIL 1	
Feature (SIL 3 and above)	**Evidence**
FMEA/Fault tree approach Common cause analysis of diverse software	

CHAPTER 5

MEETING IEC 61511

This chapter gives an overview of the requirements set out in the **process sector** specific standard IEC 61511.

The standard was issued at the beginning of 2003 and is in three parts:

Part 1	The normative standard
Part 2	Informative guidance on Part 1
Part 3	Informative guidance on hazard and risk analysis

Part 1 of the standard covers the life-cycle including:

Management of Functional Safety
Hazard and Risk Analysis
Safety Instrumented Systems (SIS) Design
through to
SIS decommissioning

The standard is intended for the activities of SIS **system level designers, integrators and users** in the process industry.

Component level **product suppliers**, such as field devices and logic solvers, are referred back to IEC 61508 as is everyone in the case of SIL 4.

Part 2 gives general guidance to the use of Part 1 on a paragraph-by-paragraph basis.

Part 3 gives more detailed guidance on targeting the Safety Integrity Levels and has a number of appendixes covering both quantitative and qualitative methods.

Since the standard is only aiming at the integration level of the SIS, rather than the individual elements, the requirements for design and development of the SIS (covered by Parts 2 and 3 of IEC 61508) have been significantly simplified. Hardware design has been replaced by a top-level set of straightforward requirements, such as '*unless otherwise justified the system shall include a manual shutdown mechanism which bypasses the logic solver*'. The software requirements are restricted to the applications software using either limited variability languages or fixed programs. Thus, the software requirement tables that are given in Part 3 of IEC 61508 have been expressed in textual terms using the requirements for SIL 3 but, in general, confined to the 'HR' items and using engineering judgement on the suitability at the applications level. For applications software using full variability languages the user is referred to IEC 61508.

The techniques and measures detailed within IEC 61511, and hence this chapter, are suitable for the development and modification of the E/E/PE system architecture and software using Limited Variability Languages up to SIL 3 rated safety functions. Unless specifically identified the same techniques and measures will be used for SILs 1, 2 and 3.

Where a project involves the development and modification of a system architecture and application software for SIL 4, or the use of Full Variability Languages for applications software (or the development of a sub-system product) then IEC 61508 should be used.

Figure 5.1 shows the relationship between 61511 and 61508.

5.1 Organising and managing the life-cycle

The requirements for the management of functional safety and life-cycle activities are basically the same as given in IEC 61508 and therefore covered by the preceding chapters. The life-cycle is required to be included in the project Quality and Safety Plan.

Each phase of the life-cycle needs to be verified for:

- Adequacy of the outputs from the phase against the requirements stated for that particular phase
- Adequacy of the review, inspection and/or testing coverage of the outputs
- Compatibility between the outputs generated at different life-cycle phases

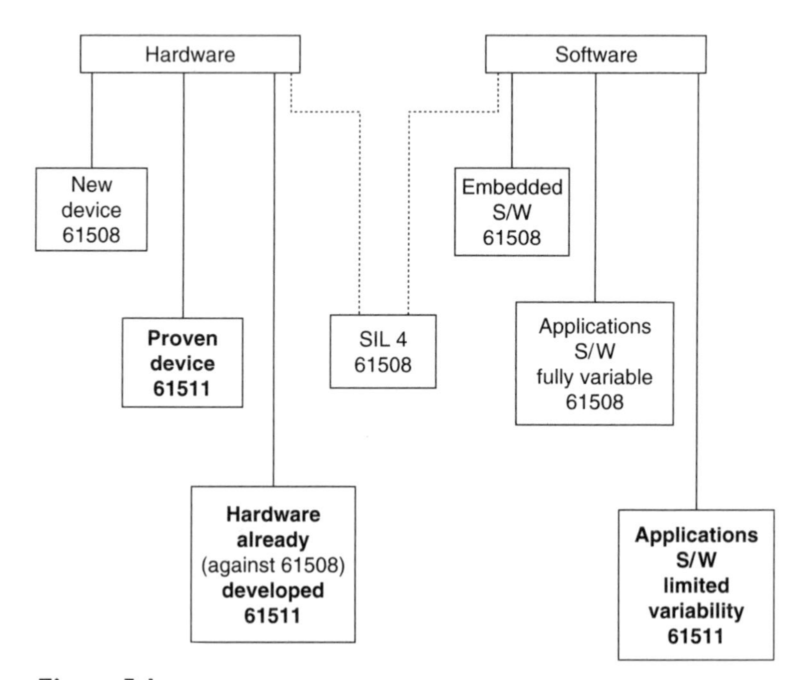

Figure 5.1
IEC 61511 versus IEC 61508

- Correctness of any data generated
- Performance of the installed safety-related system in terms of both systematic and hardware failures compared to those assumed in the design phase
- Actual demand rate on the safety system compared with the original assessment

If at any stage of the life-cycle, a change is required which affects an earlier life-cycle phase, then that earlier phase (and the following phases) need to be re-examined and, if changes are required, repeated and reverified.

The assessment team should include at least one senior, competent person not involved in the project design. All assessments will be identified in the safety plan and, typically, should be done:

- After the hazard and risk assessment
- After the design of the safety-related system
- After the installation and development of the operation/ maintenance procedures

- After gaining operational/maintenance experience
- After any changes to plant or safety system

The requirement to perform a hazard and risk analysis is basically the same as for IEC 61508 but with additional guidance being given in Part 3.

Part 1 of 61511 describes the typical layers of risk reduction (namely Control and monitoring, Prevention, Mitigation, Plant emergency response and Community emergency response). All of these should be considered as means of reducing risk and their contributing factors need to be considered in deriving the safety requirement for any safety instrumented system, which forms part of the PREVENTION layer.

Part 3 gives examples of numerical approaches, a number of risk graphs and LOPA (as mentioned in Section 2.2.1 of Chapter 2).

5.2 Requirements involving the specification

The system Functional Design Specification (FDS) will address the PES system architecture and application software requirements. The following need to be included:

- Definition of safety functions, including SIL targets
- Requirements to minimise common cause failures
- Modes of operation, with the assumed demand rate on the system
- A description of process measurements (with their trip points) and output actions
- Sub-system and component selection referencing evidence of suitability at the specified SIL requirement
- Hardware architecture
- Hardware fault tolerance
- Capacity and response time performance that is sufficient to maintain plant safety
- Environmental performance
- Power supply requirements and protection (e.g. under/overvoltage) monitoring

- Operator interfaces and their operability including:
  - Indication of automatic action
  - Indication of overrides/bypasses
  - Indication of alarm and fault status
- Procedures for non-steady state of both the plant and safety system, i.e. startup, resets etc.
- Action taken on bad process variables (e.g. sensor value out of range, detected open circuit, detected short circuit)
- Software self-monitoring, if not part of the system level software
- Proof tests and diagnostic test requirements for the logic unit and field devices
- Repair times and action required on detection of a fault to maintain the plant in a safe state
- Identification of any sub-components that need to survive an accident event, e.g. output valve that needs to survive a fire
- Design to take into account human capability for both the operator and maintenance staff
- Manual means of independently (to the logic unit) operating the final element should be specified unless otherwise justified by the safety requirements

Safety functions will be described using semi-formal methods such as Cause and Effect Charts, Logic Diagrams or Sequence Charts.

5.3 Requirements for design and development

5.3.1 Selection of components and sub-systems

Components and sub-systems for use in safety instrumented systems should either be in accordance with IEC 61508 or meet the requirements for selection based on prior use given in IEC 61511 as summarised below.

The Standard gives guidance on the use of field devices and non-PE logic solvers for up to SIL 3 safety functions using proven-in-use justification. For PE logic solvers, such as standard PLCs, guidance on the use for up to SIL 2 safety functions is given using proven-in-use justification.

For non-PE logic solvers and field devices (no software, up to SIL 3) the requirements are based on:

- Manufacturer's quality and configuration management
- Adequate identification and specification
- Demonstration of adequate performance in similar operation
- Volume of experience

For field devices (FPL software, up to SIL 3) the requirements are based on:

- As above
- Consider I/P and O/P characteristics; mode of use; function and configuration
- For SIL 3 formal assessment required

For logic solvers (up to SIL 2) the requirements are based on:

- As for field devices
- Experience taking account of SIL, complexity, and functionality
- Understanding unsafe failure modes
- Use of configuration that addresses failure modes
- Software has history in safety-related application
- Protection against unauthorised/unintended modification
- Formal assessment for SIL 2 applications

5.3.2 Architecture (i.e. safe failure fraction)

The standard provides two minimum configuration tables, one for the PE logic solvers, the other for non-PE logic solvers and field devices. Unfortunately, both tables are **formatted differently to the IEC 61508 table** and assume type B sub-systems only (i.e. the typical sub-systems used in the process industry are not assumed to be simple devices and/or do not have good reliability data. For the PE logic solvers the maximum practical SFF is assumed to be between 90% and 99%. For the non-PE logic solvers and field devices an SFF of between 60% and 90% is assumed. At any time the table in IEC 61508 can be used (see Chapter 3.3.2). For interest the 61511 version is shown below.

PE/Logic SIL	**SFF <60%**	**SFF 60–90%**	**SFF >90%**	
1	1	0	0	
2	2	1	0	Type B
3	3	2	1	
4	See IEC 61508 Part 2 Table 2 (Chapter 3)			

Non-PE SIL	**SFF <60%**	**SFF 60–90%**	**SFF >90%**	
1	***0***	0		
2	***1***	1		Type B shown thus
3	***2***	2		***Type A (simple) shown thus***
4	See IEC 61508 Part 2 Table 3 (Chapter 3)			

The 0 represents simplex. The 1 represents *m* out of *m* + 1 etc.

5.3.3 Predict the random hardware failures

Random hardware failures will be predicted as discussed in Chapter 3.

5.3.4 Software

(a) Requirements

The application software architecture needs to be consistent with the hardware architecture and satisfy the safety-integrity requirements.

The application software design shall:

- Be traceable to the requirements
- Be testable
- Include data integrity and reasonableness checks as appropriate
  - Communication link end to end checks (rolling number checks)

- Range checking on analogue sensor inputs (under- and over-range)
- Bounds checking on data parameters (i.e. have minimum size and complexity)

(b) Software library modules
Previously developed application software library modules should be used where applicable.

(c) Software design specification
A software design specification will be provided detailing:

- Software architecture
- The specification for all software modules and a description of connections and interactions
- The order of logical processing
- Any non-safety-related function that is not designed in accordance with this procedure and evidence that it cannot affect correct operation of the safety-related function

A competent person, as detailed in the Quality and Safety Plan, will approve the software design specification.

(d) Code
The application code will:

- Conform to an application specific coding standard
- Conform to the safety manual for the logic solver where appropriate
- Be subject to code inspection

(e) Programming support tools
The standard programming support tools provided by the logic solver manufacturer will be utilised together with the appropriate safety manual.

5.4 Integration and test (referred to as verification)

The following minimum verification activities need to be applied:

- Design review on completion of each life-cycle phase
- Individual software module test
- Integrated software module test

Factory acceptance testing will be carried out to ensure that the logic solver and associated software together satisfy the requirements defined in the safety requirements specifications. This will include:

- Functional test of all safety functions in accordance with the safety requirements
  - Inputs selected to exercise all specified functional cases
  - Input error handling
- Module and system level fault insertion
- System response times including 'flood alarm' conditions

5.5 Validation (meaning overall acceptance test and close-out of actions)

System validation will be provided by a factory acceptance test and a close-out audit at the completion of the project.

The complete system shall be validated by inspection and testing that the installed system meets all the requirements, that adequate testing and records have been completed for each stage of the life-cycle and that any deviations have been adequately addressed and closed out. As part of this system validation the application software validation, if applicable, needs to be closed out.

5.6 Modifications

Modifications will be carried out using the same techniques and procedures as used in the development of the original code. Change proposals will be positively identified, by the Project Safety Authority, as safety related or non-safety related. All safety-related change proposals will involve a design review, including an impact analysis, before approval.

5.7 Installation and commissioning

An installation and commissioning plan will be produced which prepares the system for final system validation. As a minimum the plan should include checking for completeness (earthing, energy sources, instrument calibration, field devices operation, logic solver operation and all operational interfaces). Records of all the testing results shall be kept and any deviations evaluated by a competent person.

5.8 Operations and maintenance

The object of this phase of the life-cycle is to ensure that the required SIL of each safety function is maintained and to ensure that the hazard demand rate on the safety system and the availability of the safety system are consistent with the original design assumptions. If there are any significant increases in hazard demand rate or decreases in the safety system availability between the design assumptions and those found in the operation of the plant which would compromise the plant safety targets then changes to the safety system will have to be made in order to maintain the plant safety.

The operation and maintenance planning needs to address:

- Routine and abnormal operation activities
- Proof testing and repair maintenance activities
- Procedures, measures and techniques to be used
- Recording of adherence to the procedures
- Recording of all demands on the safety system along with its performance to these demands
- Recording of all failures of the safety system
- Competency of all personnel
- Training of all personnel

5.9 Presenting the results

In order to justify that the SIL requirements have been correctly selected and satisfied, it is necessary to provide a documented assessment.

> Chapters 3 and 4 provided Conformance Demonstration Templates comprising simplified tables for demonstrating conformance in the case of straightforward applications designs.
>
> These may be used for demonstrating compliance to IEC 61511, using the SIL 3 levels of the Chapter 4 tables. However, some of the items may not be applicable at this application type level.

PART C

THE QUANTITATIVE ASSESSMENT

Chapters 6 and 7 explain the techniques of quantified reliability prediction and are condensed from *Reliability Maintainability and Risk*, 6th Edition, David J Smith, Butterworth-Heinemann (ISBN 0 7506 5168 7).

These two chapters largely concern random hardware failures. They go beyond IEC 61508 in providing a calibrated common cause failure model and a method of applying confidence limits to reliability predictions.

CHAPTER 6

RELIABILITY MODELLING TECHNIQUES

6.1 Failure rate and unvailability

In Chapter 1, we saw that both failure rate (λ) and probability of failure on demand (PFD) are parameters of interest. Since unavailability is the probability of being failed at a randomly chosen moment then it is the same as the probability of failure on demand. PFD is dimensionless and is given by:

$$\begin{aligned} \text{PFD} &= \text{UNAVAILABILITY} \\ &= (\lambda\ \text{MDT})/(1 + \lambda\ \text{MDT}) \cong (\lambda\ \text{MDT}) \end{aligned}$$

where λ is failure rate and MDT is the mean down time (in consistent units). Usually λ MDT $\ll 1$. For revealed failures the MDT consists of the active mean time to repair (MTTR) *plus* any logistic delays (e.g. travel, site access, spares procurement, administration).

For unrevealed failures the MDT is related to the proof-test interval (T), *plus* the active MTTR, *plus* any logistic delays. The way in which failure is defined determines, to some extent, what is included in the down time. If the unavailability of a process is confined to failures whilst production is in progress then outage due to scheduled preventive maintenance is not included in the definition of failure. However, the definition of dormant failures of redundant units affects the overall unavailability (as calculated by the equations in the next section).

6.2 Creating a reliability model

For any reliability assessment to be meaningful it is vital to address a specific system failure mode. Predicting the 'spurious shutdown' frequency of a safety (shutdown) system will involve a different logic model and different failure rates from predicting the probability of 'failure to respond'. To illustrate this, consider the case of a duplicated shutdown system whereby the voting arrangement is such that whichever sub-system recognises a valid shutdown requirement then shutdown takes place (in other words '1 out of 2' voting).

When modelling the 'failure to respond' event the '1 out of 2' arrangement represents redundancy and the two sub-systems are said to be 'parallel' in that they both need to fail to cause the event. Furthermore, the component failure rates used will be those which lead to ignoring a genuine signal. On the other hand, if we choose to model the 'spurious shutdown' event the position is reversed and the sub-systems are seen to be 'series' in that either failure is sufficient to cause the event. Now the component failure rates will be of the modes which lead to a spurious signal.

The two most commonly used modelling methods are reliability block diagram analysis and fault tree analysis.

6.2.1 Block diagram analysis

Using the above example of a shutdown system, the concept of a series reliability block diagram (RBD) applies to the 'spurious shutdown' case.

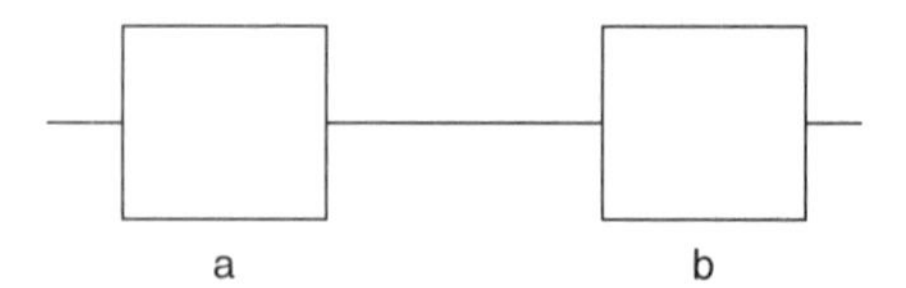

Figure 6.1
Series RBD

The two sub-systems (a and b) are described as being 'in series' since either failure causes the system failure in question.

The mathematics of this arrangement is simple. We *add* the failure rates (or unavailabilities) of series items. Thus:

$\lambda(\text{system}) = \lambda(\text{a}) + \lambda(\text{b})$

$\text{PFD}(\text{system}) = \text{PFD}(\text{a}) + \text{PFD}(\text{b})$

However, the 'failure to respond' case is represented by the parallel block diagram model as follows:

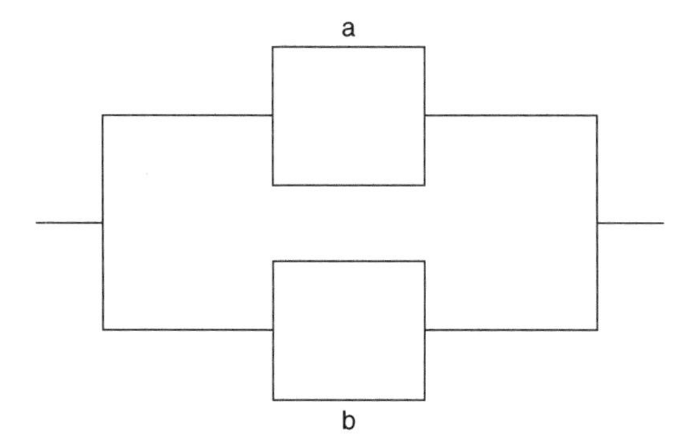

Figure 6.2
Parallel (Redundant) RBD

The mathematics is dealt with in *Reliability Maintainability and Risk*. However, the traditional results given in *Reliability Maintainability and Risk* and the majority of textbooks and standards have been challenged by K G L Simpson. It is now generally acknowledged that the traditional Markov model does not correctly represent the normal repair activities for redundant systems. The *Journal of the Safety and Reliability Society*, Vol 22, No 2, Summer 2002, published a paper by W G Gulland which agreed with those findings. The results are summarised here to enable the failure rates and unavailabilities of redundant configurations to be calculated.

The suitability of Markov modelling of redundant repairable systems has been questioned by a number of people. Ken Simpson has studied a range of redundant systems and applied various different techniques to calculate system failure rates and unavailabilities. When comparing the results from the different techniques there is good agreement with the exception of conventional Markov modelling which shows a pessimistic difference of 2:1 for a 1oo2 and up to 24:1 for 1oo4 voted systems.

This is because repair of multiple failures is not a Markov process (namely that the probability of being in a state can be determined solely from knowledge of the previous state).

For a redundant repairable system without a dedicated repair crew per equipment the transition from a multiple failure state does not depend on the repair of the last failure (as it should for the process to be applicable to a Markov model) but on the continued repair of the previous failure. For this reason a Markov model of this system is pessimistic as it underestimates the transition rate from the failed state. It is as if the repair crew abandon the earlier repair to carry out the repair of the latest failure.

With a dedicated repair crew per equipment the repair of the last failure is independent of preceding failures and the process is a Markov one. The calculations give the correct answer, which in real life situations is not a practicable situation.

For a redundant repairable system with faults detected at periodic inspection for failed items the process is also not a Markov one as the transition rate from the failed state (multiple failures) is a function of the time spent in the previous state (only one item failed). The KGLS paper (above) should be consulted for a deeper understanding.

It is worth mentioning that, as with all redundant systems, the total system failure rate (or PFD) will be dominated by the effect of common cause failure dealt with later in this chapter. Tables 6.1 and 6.2 provide the failure rate and unavailability equations for simplex and parallel (redundant) identical subsystems for revealed failures having a mean downtime of MDT.

Table 6.1 System failure rates (revealed)

Number of units					
	1	λ			
	2	$2\lambda^2$MDT	2λ		
	3	$3\lambda^3$MDT^2	$6\lambda^2$MDT	3λ	
	4	$4\lambda^4$MDT^3	$12\lambda^3$MDT^2	$12\lambda^2$MDT	4λ
		1	2	3	4
		Number required to operate			

Table 6.2 System unavailabilities (revealed)

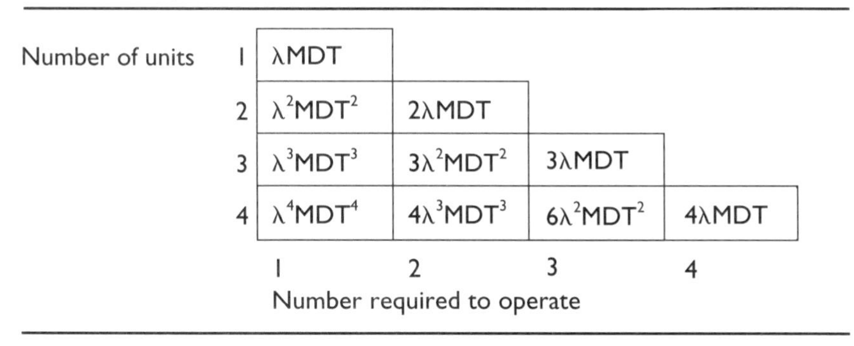

Number of units	1	2	3	4
1	λMDT			
2	$\lambda^2 MDT^2$	2λMDT		
3	$\lambda^3 MDT^3$	$3\lambda^2 MDT^2$	3λMDT	
4	$\lambda^4 MDT^4$	$4\lambda^3 MDT^3$	$6\lambda^2 MDT^2$	4λMDT

Number required to operate

Unrevealed failures will eventually be revealed by some form of auto-test or proof-test. Whether manually scheduled or automatically initiated (e.g. auto-test using programmable logic) there will be a proof-test interval, T.

Tables 6.3 and 6.4 provide the failure rate and unavailability equations for simplex and parallel (redundant) identical

Table 6.3 Failure rates (unrevealed)

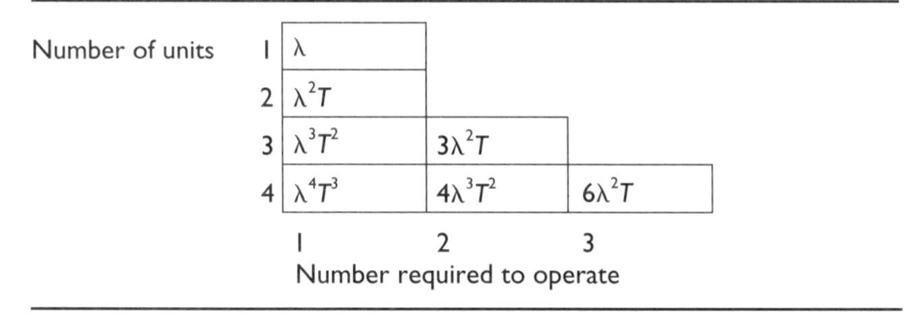

Number of units	1	2	3
1	λ		
2	$\lambda^2 T$		
3	$\lambda^3 T^2$	$3\lambda^2 T$	
4	$\lambda^4 T^3$	$4\lambda^3 T^2$	$6\lambda^2 T$

Number required to operate

Table 6.4 Unavailabilities (unrevealed)

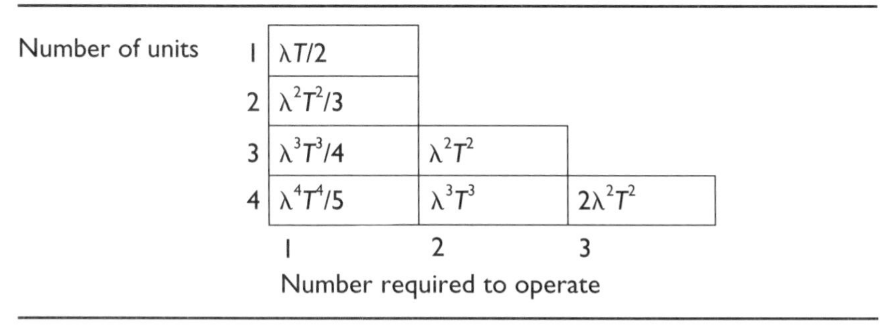

Number of units	1	2	3
1	$\lambda T/2$		
2	$\lambda^2 T^2/3$		
3	$\lambda^3 T^3/4$	$\lambda^2 T^2$	
4	$\lambda^4 T^4/5$	$\lambda^3 T^3$	$2\lambda^2 T^2$

Number required to operate

sub-systems for unrevealed failures having a proof-test interval T. The MTTR is assumed to be negligible compared with T.

6.2.2 Common cause failure (CCF)

Whereas simple models of redundancy assume that failures are both random and independent, common cause failure (CCF) modelling takes account of failures which are linked, due to some dependency, and therefore occur simultaneously or, at least, within a sufficiently short interval as to be perceived as simultaneous.

Two examples are:

(a) The presence of water vapour in gas causing two valves to seize due to icing. In this case the interval between the two failures might be in the order of days. However, if the proof-test interval for this dormant failure is two months then the two failures will, to all intents and purposes, be simultaneous.
(b) Inadequately rated rectifying diodes on identical twin printed circuit boards failing simultaneously due to a voltage transient.

Typically, causes arise from

(a) Requirements: incomplete or conflicting.
(b) Design: common power supplies, software, emc, noise.
(c) Manufacturing: batch related component deficiencies.
(d) Maintenance/operations: human induced or test equipment problems.
(e) Environment: temperature cycling, electrical interference etc.

Defences against CCF involve design and operating features which form the assessment criteria given in Appendix 3.

Common cause failures often dominate the unreliability of redundant systems by virtue of defeating the random coincident failure feature of redundant protection. Consider the duplicated system in Figure 6.2. The failure rate of the redundant element (in other words the coincident failures) can be calculated using the formula developed in Table 6.1, namely

$2\lambda^2$MDT. Typical figures of 10 per million hours failure rate (10^{-5} per hr) and 24 hours down time lead to a failure rate of $2 \times 10^{-10} \times 24 = 0.0048$ per million hours. However, if only one failure in 20 is of such a nature as to affect both channels and thus defeat the redundancy, it is necessary to add the series element, shown as λ_2 in Figure 6.3, whose failure rate is $5\% \times 10^{-5} = 0.5$ per million hours, being two orders more frequent. The 5%, in this example, is known as a BETA factor. The effect is to swamp the redundant part of the prediction and it is thus important to include CCF in reliability models. This sensitivity of system failure to CCF places emphasis on the credibility of CCF estimation and thus justifies efforts to improve the models.

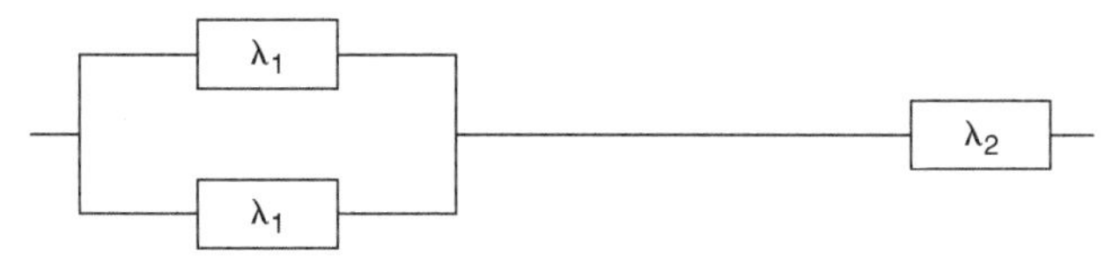

Figure 6.3
Reliability block diagram showing CCF

In Figure 6.3, (λ_1) is the failure rate of a single redundant unit and (λ_2) is the common cause failure rate such that (λ_2) = β(λ_1) for the BETA model, which assumes that a fixed proportion of the failures arise from a common cause. The contributions to BETA are split into groups of design and operating features which are believed to influence the degree of CCF. Thus the BETA multiplier is made up by adding together the contributions from each of a number of factors within each group. This Partial BETA model (as it is known) involves the following groups of factors, which represent defences against CCF:

- Similarity (Diversity between redundant units reduces CCF)
- Separation (Physical distance and barriers reduce CCF)
- Complexity (Simpler equipment is less prone to CCF)
- Analysis (FMEA and field data analysis will help to reduce CCF)

- Procedures (Control of modifications and of maintenance activities can reduce CCF)
- Training (Designers and maintainers can help to reduce CCF by understanding root causes)
- Control (Environmental controls can reduce susceptibility to CCF, e.g. weather proofing of duplicated instruments)
- Tests (Environmental tests can remove CCF prone features of the design, e.g. emc testing)

The Partial BETA model is assumed to be made up of a number of partial βs, each contributed to by the various groups of causes of CCF. β is then estimated by reviewing and scoring each of the contributing factors (e.g. diversity, separation).

The BETAPLUS model has been developed from the Partial BETA method because:

- it is objective and maximises traceability in the estimation of BETA. In other words the choice of checklist scores, when assessing the design, can be recorded and reviewed;
- it is possible for any user of the model to develop the checklists further to take account of any relevant failure causal factors that may be perceived;
- it is possible to calibrate the model against actual failure rates, albeit with very limited data;
- there is a credible relationship between the checklists and the system features being analysed. The method is thus likely to be acceptable to the non-specialist;
- the additive scoring method allows the partial contributors to β to be weighted separately;
- the β method acknowledges a direct relationship between (λ_2) and (λ_1) as depicted in Figure 6.3;
- it permits an assumed 'non-linearity' between the value of β and the scoring over the range of β.

The BETAPLUS model includes the following enhancements:

(a) CATEGORIES OF FACTORS:
Whereas existing methods rely on a single subjective judgement of score in each category, the BETAPLUS method provides specific design and operationally related questions to be answered in each category.

(b) SCORING:
The maximum score for each question has been weighted by calibrating the results of assessments against known field operational data.

(c) TAKING ACCOUNT OF DIAGNOSTIC COVERAGE:
Since CCF are not simultaneous, an increase in auto-test or proof-test frequency will reduce β since the failures may not occur at precisely the same moment.

(d) SUB-DIVIDING THE CHECKLISTS ACCORDING TO THE EFFECT OF DIAGNOSTICS:
Two columns are used for the checklist scores. Column (A) contains the scores for those features of CCF protection which are perceived as being enhanced by an increase in diagnostic frequency. Column (B), however, contains the scores for those features believed not to be enhanced by an improvement in diagnostic frequency. In some cases the score has been split between the two columns, where it is thought that some, but not all, aspects of the feature are affected (see Appendix 3).

(e) ESTABLISHING A MODEL:
The model allows the scoring to be modified by the frequency and coverage of diagnostic test. The (A) column scores are modified by multiplying by a factor (C) derived from diagnostic related considerations. This (C) score is based on the diagnostic frequency and coverage. (C) is in the range 1 to 3. A factor 'S', used to derive BETA, is then estimated from the RAW SCORE:

$$S = \text{RAW SCORE} = (\Sigma A \times C) + \Sigma B$$

(f) NON-LINEARITY:
There are currently no CCF data to justify departing from the assumption that, as BETA decreases (i.e. improves), then successive improvements become proportionately harder to achieve. Thus the relationship of the BETA factor to the RAW SCORE $[(\Sigma A \times C) + \Sigma B]$ is assumed to be exponential and this non-linearity is reflected in the equation which translates the raw score into a BETA factor.

(g) EQUIPMENT TYPE:
The scoring has been developed separately for programmable and non-programmable equipment, in order to

reflect the slightly different criteria which apply to each type of equipment.

(h) CALIBRATION:
The model has been calibrated against field data.

Scoring criteria were developed to cover each of the categories (i.e. separation, diversity, complexity, assessment, procedures, competence, environmental control, environmental test). Questions have been assembled to reflect the likely features which defend against CCF. The scores were then adjusted to take account of the relative contributions to CCF in each area, as shown in the author's data. The score values have been weighted to calibrate the model against the data.

When addressing each question (in Appendix 3) a score, less than the maximum of 100%, may be entered. For example, in the first question, if the judgement is that only 50% of the cables are separated then 50% of the maximum scores (15 and 52) may be entered in each of the (A) and (B) columns (7.5 and 26).

The checklists are presented in two forms (listed in Appendix 3) because the questions applicable to programmable based equipments will be slightly different to those necessary for non-programmable items (e.g. field devices and instrumentation).

The headings are:

(1) SEPARATION/SEGREGATION
(2) DIVERSITY/REDUNDANCY
(3) COMPLEXITY/DESIGN/APPLICATION/ MATURITY/EXPERIENCE
(4) ASSESSMENT/ANALYSIS and FEEDBACK OF DATA
(5) PROCEDURES/HUMAN INTERFACE
(6) COMPETENCE/TRAINING/SAFETY CULTURE
(7) ENVIRONMENTAL CONTROL
(8) ENVIRONMENTAL TESTING

Assessment of the diagnostic interval factor (C)

In order to establish the (C) score it is necessary to address the effect of diagnostic frequency. The diagnostic coverage, expressed as a percentage, is an estimate of the proportion of failures which would be detected by the proof test or auto-test. This can be estimated by judgement or, more formally, by applying FMEA at the component level to decide whether each failure would be revealed by the diagnostics.

An exponential model is used to reflect the increasing difficulty in further reducing BETA as the score increases. This is reflected in the following equation which is developed in Smith D J (2000). Developments in the use of failure rate data…':

$$\beta = 0.3 \exp(-3.4S/2624)$$

6.2.3 Fault tree analysis

Whereas the reliability block diagram provides a graphical means of expressing redundancy in terms of 'parallel' blocks, fault tree analysis expresses the same concept in terms of paths of failure. The system failure mode in question is referred to as the *top event* and the paths of the tree represent combinations of event failures leading to the *top event*. The underlying mathematics is exactly the same. Figure 6.4 shows the OR gate which is equivalent to Figure 6.1 and the AND gate which is equivalent to Figure 6.2.

Figure 6.5 shows a typical fault tree modelling the loss of fire water which arises from the failure of a pump, a motor, the detection and the combined failure of both power sources.

In order to allow for common cause failures in the fault tree model, additional gates are drawn as shown in the following examples. Figure 6.6 shows the reliability block diagram of Figure 6.3 in fault tree form. The common cause failure can be seen to defeat the redundancy by introducing an OR gate along with the redundant G1 gate.

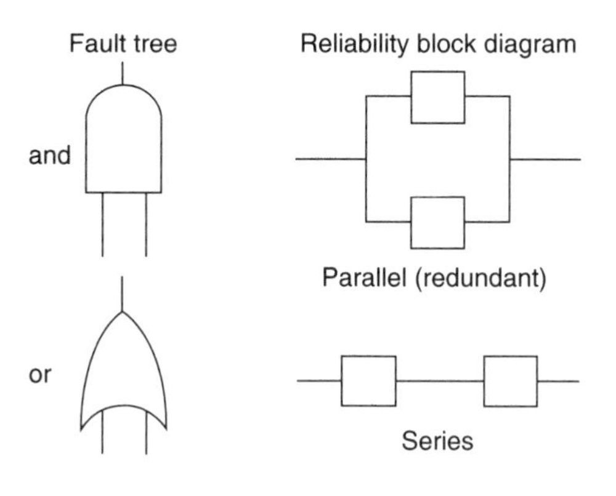

Figure 6.4
Series and parallel equivalent to OR and AND

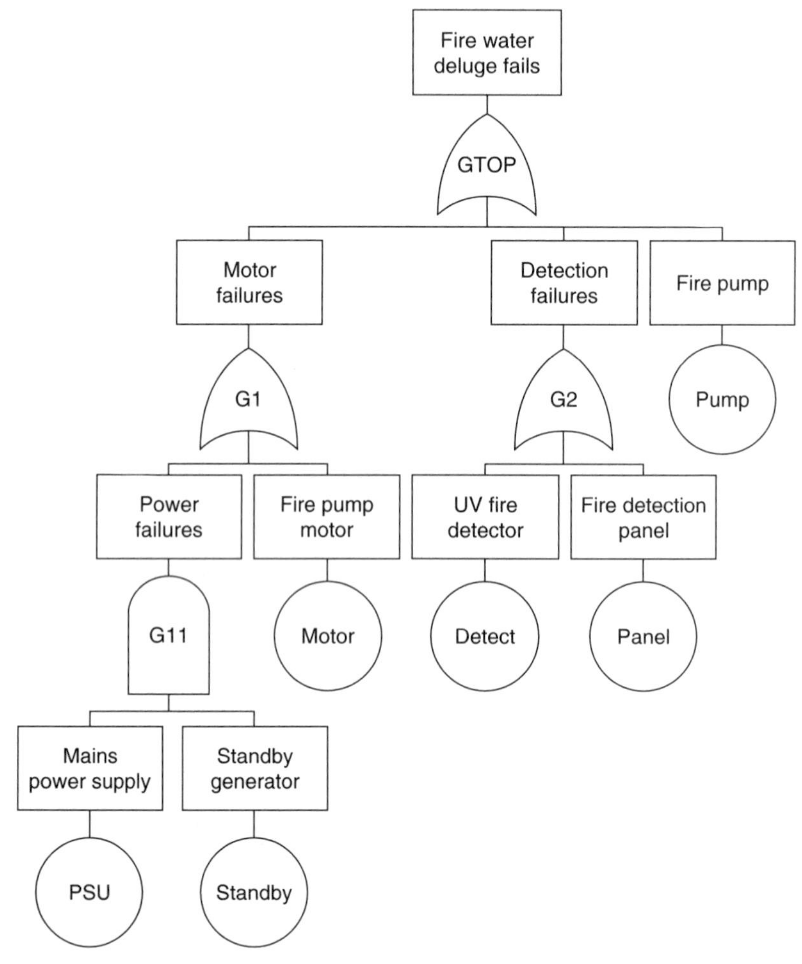

Figure 6.5
Example of a fault tree

Figure 6.7 shows another example, this time of '2 out of 3' redundancy, where a voted gate is used.

6.3 Taking account of auto-test

The mean down time (MDT) of unrevealed failures can be assessed as a fraction of the proof-test interval (i.e. for random failures, an average of half the proof-test interval as far as an individual unit is concerned) plus the actual MTTR (mean time to repair).

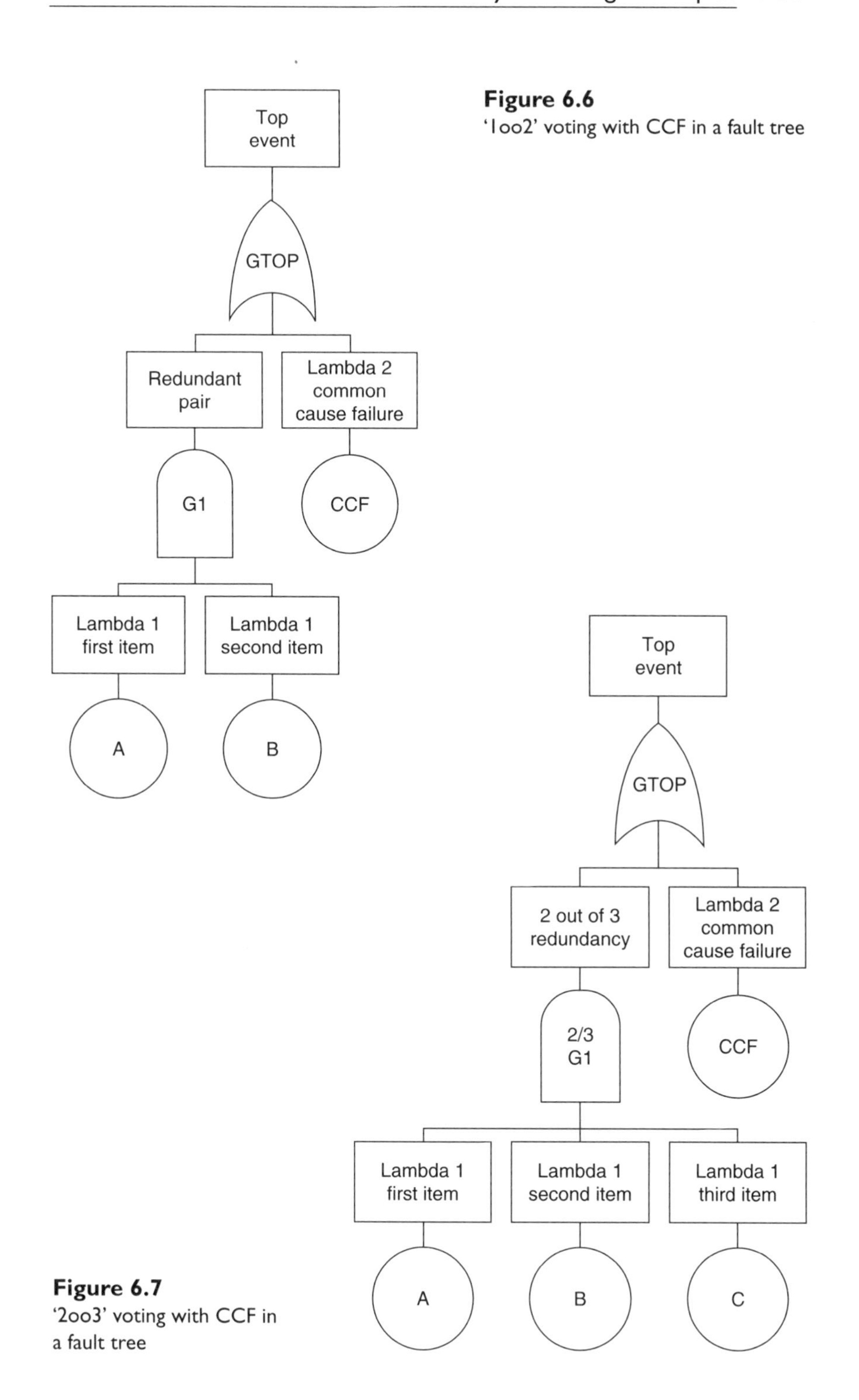

Figure 6.6
'1oo2' voting with CCF in a fault tree

Figure 6.7
'2oo3' voting with CCF in a fault tree

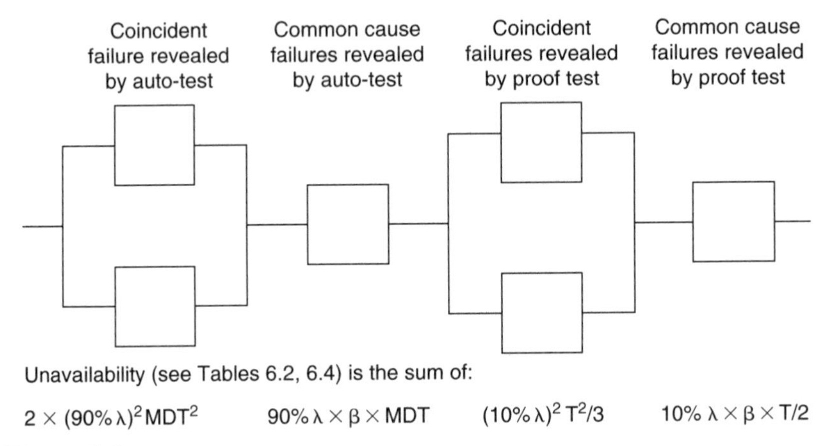

Figure 6.8
Reliability block diagram, taking account of diagnostics

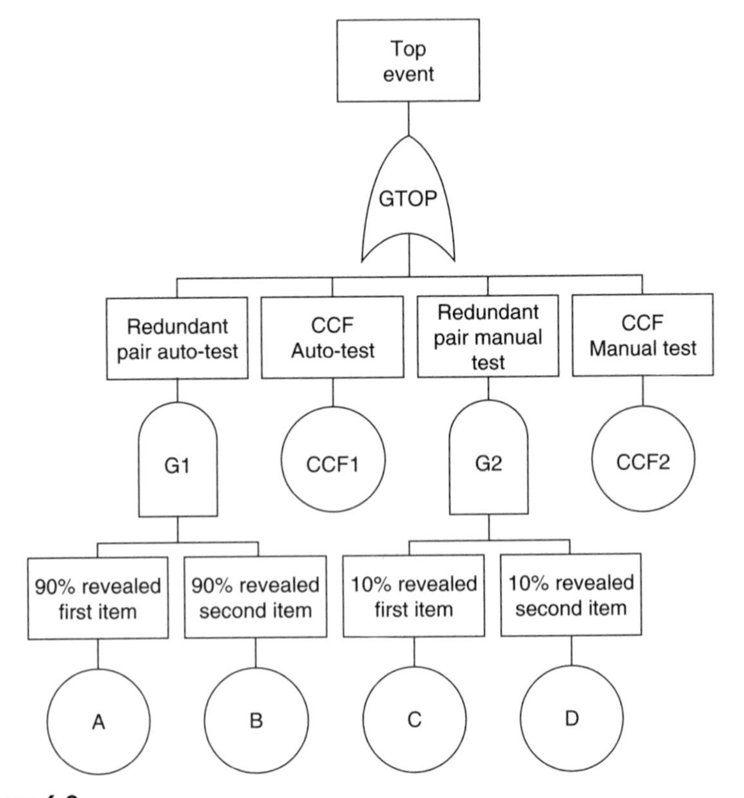

Figure 6.9
Equivalent fault tree to Figure 6.8

In many cases there is both auto-test, whereby a programmable element in the system carries out diagnostic checks to discover unrevealed failures, as well as a manual proof test. In practice the auto-test will take place at some relatively short interval (e.g. 8 minutes) and the proof test at a longer interval (e.g. 4000 hours).

The question arises as to how the reliability model takes account of the fact that failures revealed by the auto-test enjoy a shorter down time than those left for the proof test. The ratio of one to the other is a measure of the diagnostic coverage and is expressed as a percentage of failures revealed by the test.

Consider now a dual redundant configuration (voted 1 out of 2) subject to 90% auto-test and the assumption that the manual test reveals 100% of the remaining failures. The reliability block diagram needs to split the model into two parts in order to calculate separately in respect of the auto-diagnosed and manually-diagnosed failures. Figure 6.8 shows the parallel and common cause elements twice and applies the equations from Section 6.2 to each element. The failure rate of the item, for the failure mode in question, is λ. The equivalent fault tree is shown in Figure 6.9.

6.4 Human error/human factors

In addition to random coincident hardware failures, and their associated dependent failures (previous section), it is frequently necessary to include human error in a prediction model (e.g. fault tree). Specific quantification of human error factors is not a requirement of IEC 61508. However, it is required that human factors are 'considered'.

It is well known that the majority of well-known major incidents, such as Three Mile Island, Bhopal, Chernobyl, Zeebrugge, Clapham and Paddington are related to the interaction of complex systems with human beings. In short, the implication is that human error was involved, to a greater or lesser extent, in these and similar incidents. For some years there has been an interest in modelling these factors so that quantified reliability and risk assessments can take account of the contribution of human error to the system failure.

IEC 61508 (Part 1) requires the consideration of human factors at a number of places in the life-cycle. The assessment of

Part 1		
Para. 1.2	Scope	Makes some reference
Para. 1.3	Management	Reference to training
Table 1	Life-cycle	Several uses of 'to include human factors'
Para. 7.3.2.4	Definition stage	Human error to be considered
Para. 7.4 various	Hazard/Risk analysis	References to misuse and human intervention
Para. 7.6.2.2	Safety requirements allocation	Availability of skills
Paras 7.7.2.3 & 7.15.2.1	Ops & maintenance	Refers
Part 2		
Para. 7.3.2.1 (c)	Requirements	Operator interfaces
Para. 7.4	Design and development	Operator interfaces; manual means of switching to safe state; tolerance to operator error; account for human limitations
Paras 7.6.2.1 & 7.6.2.3	Ops & maintenance	Routines to maintain safe state to be defined
Para. 7.7.2.3	Validation	Includes procedures
Para. 7.8.2.1	Modification	Evaluate mods on their effect on human interaction
Part 3		
Para. 7.2.2.4	Specification	Operator interfaces
Para. 7.9.2.13 (c)	Verification	Modifiable parameters to be protected from unauthorised change

human error is therefore implied. The following summarises the main references in the Standard.

There are also references in Parts 4 to 7 and these can be found listed in a fuller treatment of this subject *Proposed Framework for Addressing Human Factors in IEC 61508*, M Carey, Amey VECTRA Ltd.

One example might be a process where there are three levels of defence against a specific hazard (e.g. over-pressure of a vessel). In this case the control valve will be regarded as the EUC. The three levels of defence are:

(1) The control system maintaining the setting of a control valve.
(2) A shutdown system operating a separate shut-off valve in response to a high pressure.
(3) Human response whereby the operator observes a high pressure reading and inhibits flow from the process.

The risk assessment would clearly need to consider how independent of each other are these three levels of protection. If the operator action (3) invokes the shutdown (2) then failure of that shutdown system will inhibit both defences. In either case the probability of operator error (failure to observe or act) is part of the quantitative assessment.

Another example might be air traffic control, where the human element is part of the safety loop rather than an additional level of protection. In this case human factors are safety-critical rather than safety-related.

Human error rate data for various forms of activity, particularly in operations and maintenance, are needed. In the early 1960s there were attempts, by UKAEA, to develop a database of human error rates and these led to models of human error whereby rates could be estimated by assessing relevant factors such as stress, training and complexity. These human error probabilities include not only simple failure to carry out a given task, but diagnostic tasks where errors in reasoning, as well as action, are involved. There is not a great deal of data available due to the following problems:

- Low probabilities require large amounts of experience in order for meaningful statistics to emerge.

- Data collection concentrates on recording the event rather than analysing the causes.
- Many large organisations have not been prepared to commit the necessary resources to collect data.

More recently interest has developed in exploring the underlying reasons, as well as probabilities, of human error. As a result there are currently several models, each developed by separate groups of analysts working in this field.

Estimation methods are described in the UKAEA document SRDA-R11, 1995. The better known are HEART (Human Error Assessment and Reduction Technique), THERP (Technique for Human Error Rate Prediction) and TESEO (Empirical Technique To Estimate Operator Errors).

For the earlier over-pressure example, failure of the operator to react to a high pressure (3) might be modelled by two of the estimation methods as follows:

'HEART' method

Basic task 'Restore system following checks' – error rate = 0.003

Modifying factors:

Few independent checks	×3	50%
No means of reversing decision	×8	25%

An algorithm is provided (not in the scope of this book).

Thus error probability $= 0.003 \times [2 \times 0.5 + 1] \times [7 \times 0.25 + 1]$
$$\mathbf{= 1.6 \times 10^{-2}}$$

'TESEO' method

Basic task 'Requires attention' – error rate = 0.01

Modifying factors:

for stress	×1
for operator	×1
for emergency	×2
for ergonomic factors	×1

Thus error probability $= 0.01 \times 1 \times 1 \times 2 \times 1$
$$\mathbf{= 2 \times 10^{-2}}$$

The two methods are in fair agreement and thus a figure of:

2×10^{-2} MIGHT BE USED FOR THE EXAMPLE

Figure 6.10 shows a fault tree for the example assuming that the human response is independent of the shutdown system.

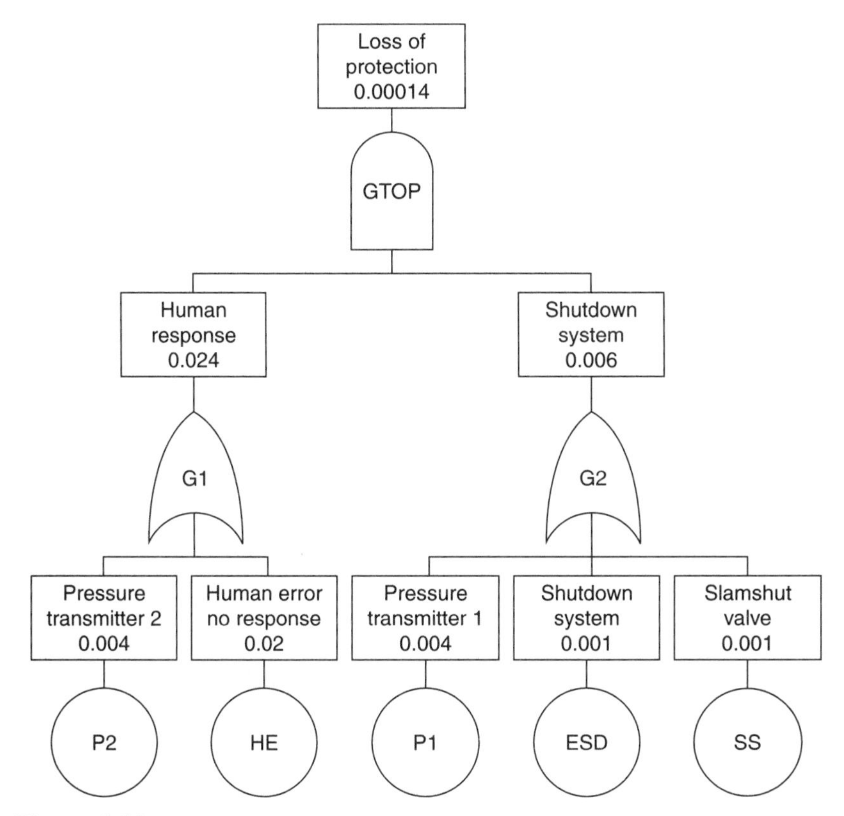

Figure 6.10
Fault tree for HEART/TESEO example

The fault tree models the failure of the two levels of protection (2) and (3). Typical (credible) probabilities of failure on demand are used for the initiating events. The human error value of 2×10^{-2} could well have been estimated as above.

Quantifying this tree would show that the overall probability of failure on demand is 1.4×10^{-4} (incidentally meeting SIL 3 quantitatively). Looking at the relative contribution of the combinations of initiating events would show that human error is involved in over 80% of the total. Thus, further consideration of human error factors would be called for.

AFTER READING CHAPTER 7 TRY THE EXERCISE AND THE EXAMPLES IN CHAPTERS 11 AND 12.

CHAPTER 7

FAILURE RATE AND MODE DATA

In order to quantify reliability models it is necessary to obtain failure rate and failure mode data.

7.1 Data accuracy

There are many collections of failure rate data compiled by defence, telecommunications, process industries, oil and gas and other organisations. Some are published Data Handbooks such as:

US MIL HANDBOOK 217 (Electronics)
CNET (French PTT) Data
HRD (Electronics, British Telecom)
RADC Non-Electronic Parts Handbook NPRD
OREDA (Offshore data)
FARADIP.THREE (Data ranges)

Some are data banks which are accessible by virtue of membership or consultancy fee such as:

SRD (Part of UKAEA) Data Bank
Technis (Tonbridge)

Some are in-house data collections which are not generally available. These occur in large industrial manufacturers and public utilities.

These data collection activities were at their peak in the 1980s but, sadly, they declined during the 1990s and many of the published sources have not been updated since that time.

Failure data are usually, unless otherwise specified, taken to refer to random failures (i.e. constant failure rates). It is important to read, carefully, any covering notes since, for a given temperature and environment, a stated component, despite the same description, may exhibit a wide range of failure rates because:

1. Some failure rate data include items replaced during preventive maintenance whereas others do not. These items should, ideally, be excluded from the data but, in practice, it is not always possible to identify them. This can affect rates by an order of magnitude.
2. Failure rates are affected by the tolerance of a design. Because definitions of failure vary, a given parametric drift may be included in one data base as a failure, but ignored in another. This will cause a variation in the values.
3. Although nominal environmental and quality assurance levels are described in some databases, the range of parameters covered by these broad descriptions is large. They represent, therefore, another source of variability.
4. Component parts are often only described by reference to their broad type (e.g. signal transformer). Data are therefore combined for a range of similar devices rather than being separately grouped, thus widening the range of values. Furthermore, different failure modes are often mixed together in the data.
5. The degree of data screening will affect the relative numbers of intrinsic and induced failures in the quoted failure rate.
6. Reliability growth occurs where field experience is used to enhance reliability as a result of modifications. This will influence the failure rate data.
7. Trial and error replacement is sometimes used as a means of diagnosis and this can artificially inflate failure rate data.
8. Some data record undiagnosed incidents and 'no fault found' visits. If these are included in the statistics as faults, then failure rates can be inflated.

Quoted failure rates are therefore influenced by the way they are interpreted by an analyst and can span one or two orders of magnitude as a result of different combinations of the above factors. **Prediction calculations were explained in Chapter 6**

and it will be seen that the relevance of failure rate data is more important than refinements in the statistics of the calculation. The data sources described in Section 7.2 can at least be subdivided into 'Site specific', 'Industry specific' and 'Generic' and work has shown (Smith D J, 2000 *Developments in the use of failure rate data*) that the more specific the data source the greater the confidence in the prediction.

Failure rates are often tabulated, for a given component type, against ambient temperature and the ratio of applied to rated stress (power or voltage). Data are presented in one of two forms:

1. *Tables:* Lists of failure rates, with or without multiplying factors, for such parameters as quality and environment.
2. *Models:* Obtained by regression analysis of the data. These are presented in the form of equations which yield a failure rate as a result of inserting the device parameters into the appropriate expression.

Because of the large number of variables involved in describing microelectronic devices, data are often expressed in the form of models. These regression equations (WHICH GIVE A TOTALLY MISLEADING IMPRESSION OF PRECISION) involve some or all of the following:

- Complexity (number of gates, bits, equivalent number of transistors).
- Number of pins.
- Junction temperature.
- Package (ceramic and plastic packages).
- Technology (CMOS, NMOS, bipolar, etc.).
- Type (memory, random LSI, analogue, etc.).
- Voltage or power loading.
- Quality level (affected by screening and burn-in).
- Environment.
- Length of time in manufacture.

Although empirical relationships have been established relating certain device failure rates to specific stresses, such as voltage and temperature, no precise formula exists which links specific environments to failure rates. The permutation of different values of environmental factors is immense. General adjustment (multiplying) factors have been evolved and these are often used to scale up basic failure rates to particular environmental conditions.

Because Failure Rate is, probably, the least precise engineering parameter, it is important to bear in mind the limitations of a Reliability prediction. The work mentioned above (Smith, 2000) makes it possible to **express predictions using confidence intervals**. The resulting MTBF, availability (or other parameter), should not be taken as an absolute value but rather as a general guide to the design reliability. Within the prediction, however, the relative percentages of contribution to the total failure rate are of a better accuracy and provide a valuable tool in design analysis.

Owing to the differences between data sources, comparisons of reliability should always involve the same data source in each prediction.

For any reliability assessment to be meaningful it must address a specific system failure mode. To predict that a safety (shutdown) system will fail at a rate of, say, once per annum is, on its own, saying very little. It might be that 90% of the failures lead to a spurious shutdown and 10% to a failure to respond. If, on the other hand, the ratios were to be reversed then the picture would be quite different.

The failure rates, mean times between failures or availabilities must therefore be assessed for defined failure types (modes). In order to achieve this, the appropriate component level failure modes must be applied to the prediction models which were described in Chapter 6. Component failure mode data are sparse but a few of the sources do contain some information. The following sections indicate where this is the case.

7.2 Sources of data

Sources of failure rate and failure mode data can be classified as:

1. SITE SPECIFIC
   Failure rate data which have been collected from similar equipment being used on very similar sites (e.g. two or more gas compression sites where environment, operating methods, maintenance strategy and equipment are largely the same). Another example would be the use of failure rate

data from a flow corrector used throughout a specific distribution network. These data might be applied to the RAMS (reliability, availability, maintainability, safety) prediction for a new design of circuitry for the same application.

2. INDUSTRY SPECIFIC
   An example would be the use of the OREDA offshore failure rate data book for a RAMS prediction of a proposed offshore process package.

3. GENERIC
   A generic data source combines a large number of applications and sources (e.g. FARADIP.THREE).

As has already been emphasised, predictions require failure rates for specific modes of failure (e.g. open circuit, signal high, valve closes). Some, but unfortunately only a few, data sources contain specific failure mode percentages. Mean time to repair data is even more sparse although the OREDA data base is very informative in this respect.

The following are the more widely used sources:

7.2.1 Electronic failure rates

1. *US Military Handbook 217* (Generic, no failure modes)
2. *HRD5 Handbook of Reliability Data for Electronic Components Used in Telecommunications Systems* (Industry specific, no failure modes)
3. *Recueil de Donnés de Fiabilité du CNET* (Industry specific, no failure modes)
4. *BELLCORE (Reliability Prediction Procedure for Electronic Equipment) TR-NWT-000332 Issue 5 1995* (Industry specific, no failure modes)
5. *Electronic data NOT available for purchase*

A number of companies maintain failure rate data banks including Nippon Telephone Corporation (Japan), Ericson (Sweden), and Thomson CSF (France) but these data are not generally available outside the organisations.

7.2.2 Other general data collections

1. *Nonelectronic Parts Reliability Data Book – NPRD* (Generic, some failure modes)

2. *OREDA – Offshore Reliability Data* (1984/92/95/97) (Industry specific, detailed failure modes, mean times to repair)

3. *FARADIP.THREE* (the author) (Industry and generic, many failure modes, some repair times)

4. *UKAEA* (Industry and generic, many failure modes)

5. *Sources of nuclear generation data* (Industry specific) In the UKAEA documents, above, there are some nuclear data, as in NNC (National Nuclear Corporation) although this may not be openly available. In the USA, Appendix III of the WASH 1400 study provided much of the data frequently referred to and includes failure rate ranges, event probabilities, human error rates and some common cause information. The IEEE Standard IEEE 500 also contains failure rates and restoration times. In addition there is NUCLARR (Nuclear Computerised Library for Assessing Reliability) which is a PC-based package developed for the Nuclear Regulatory Commission and containing component failure rates and some human error data. Another US source is the NUREG publication. Some of the EPRI data are related to nuclear plant. In France, Electricity de France provides the EIReDA mechanical and electrical failure rate data base which is available for sale. In Sweden the TBook provides data on components in Nordic Nuclear Power Plants.

6. *US sources of power generation data* (Industry specific) The EPRI (Electrical Power Research Institute) of GE Co, New York data scheme is largely gas turbine generation failure data in the USA. There is also the GADS (Generating Availability Data System) operated by NERC (North American Electric Reliability Council). They produce annual statistical summaries based on experience from power stations in USA and Canada.

7. *SINTEF* (Industry specific)
   SINTEF is the Foundation for Scientific and Industrial Research at the Norwegian Institute of Technology. They produce a number of reliability handbooks which include failure rate data for various items of process equipment.

8. *Data not available for purchase*
   Many companies (e.g. Siemens), and for that matter firms of RAMS consultants (e.g. RM Consultants Ltd) maintain failure rate data but only for use by that organisation.

7.2.3 Some older sources

A number of sources have been much used and are still frequently referred to. They are, however, somewhat dated but are listed here for completeness.

Reliability Prediction Manual for Guided Weapon Systems (UK MOD) – DX99/013–100
Reliability Prediction Manual for Military Avionics (UK MOD) – RSRE250
UK Military Standard 00–41
Electronic Reliability Data – INSPEC/NCSR (1981)
Green and Bourne (book), *Reliability Technology,* Wiley 1972
Frank Lees (book), *Loss Prevention in the Process Industries,* Butterworth-Heinemann 1996.

7.3 Data ranges and confidence levels

For some components there is fairly close agreement between the sources and in other cases there is a wide range, the reasons for which were summarised above. For this reason predictions are subject to wide tolerances.

The ratio of predicted failure rate (or system unavailability) to field failure rate (or system unavailability) was calculated for each of 44 examples and the results (see Smith D J, 2000) were classified under the three categories described in Section 7.2, namely:

1. *Predictions using site specific data.*
2. *Predictions using industry specific data.*
3. *Predictions using generic data.*

The results are:

1. For a prediction using site specific data

One can be this confident	*That the eventual field failure rate will be BETTER than:*
95%	3½ times the predicted
90%	2½ times the predicted
60%	1½ times the predicted

2. For a prediction using industry specific data

One can be this confident	*That the eventual field failure rate will be BETTER than:*
95%	5 times the predicted
90%	4 times the predicted
60%	2½ times the predicted

3. For a prediction using generic data

One can be this confident	*That the eventual field failure rate will be BETTER than:*
95%	8 times the predicted
90%	6 times the predicted
60%	3 times the predicted

Additional evidence in support of the 8:1 range is provided from the FARADIP.THREE data bank which shows an average of 7:1 across the ranges.

The FARADIP.THREE data base was created to show the ranges of failure rate for most component types. This database, currently version 4.1 in 2000, is a summary of most of the other databases and shows, for each component, the range of failure rate values which is to be found from them. Where a value in the range tends to predominate then this is indicated. Failure mode percentages are also included. It is available on disk from Technis at 26 Orchard Drive, Tonbridge, Kent TN10 4LG, UK and includes:

Discrete

Diodes
Opto-electronics
Lamps and displays
Crystals
Tubes

Passive
Capacitors
Resistors
Inductive
Microwave
Instruments and Analysers
Analysers
Fire and gas detection
Meters
Flow instruments
Pressure instruments
Level instruments
Temperature instruments
Connection
Connections and connectors
Switches and breakers
PCBs cables and leads
Electromechanical
Relays and solenoids
Rotating machinery (fans, motors, engines)
Power
Cells and chargers
Supplies and transformers
Mechanical
Pumps
Valves and parts
Bearings
Miscellaneous
Pneumatics
Hydraulics
Computers, data processing and communications
Alarms, fire protection, arresters and fuses

7.4 Conclusions

The use of stress-related regression models implies an unjustified precision in estimating the failure rate parameter.

Site specific data should be used in preference to industry specific data which, in turn, should be used in preference to generic data.

Predictions should be expressed in confidence limit terms using the above information.

In practice, failure rate is a system level effect. It is closely related to, but not entirely explained by, component failure. A significant proportion of failures encountered with modern electronic systems are not the direct result of parts failures but of more complex interactions within the system. The reason for this lack of precise mapping arises from such effects as human factors, software, environmental interference, inter-related component drift and circuit design tolerance.

The primary benefit to be derived from reliability and safety engineering is the reliability and integrity growth which arises from ongoing analysis and follow-up as well as from the corrective actions brought about by failure analysis. Reliability prediction, based on the manipulation of failure rate data, involves so many potential parameters that a valid repeatable model for failure rate estimation is not possible. Thus, failure rate is the least accurate of engineering parameters and prediction from past data should be carried out either:

- As an indicator of the approximate level of reliability of which the design is capable, given reliability growth in the field.
- To provide relative comparisons in order to make engineering decisions concerning optimum redundancy.
- As a contractual requirement.
- In response to safety-integrity requirements.

It should not be regarded as an exact indicator of future field reliability as a result of which highly precise prediction methods are often, by reason of poor data, not justified.

NOW TRY THE EXERCISE AND THE EXAMPLES IN CHAPTERS 11 AND 12

PART D

RELATED ISSUES

In this section we will make some comments (in Chapter 8) on Part 6 of the Standard, review other standards and guidance (in Chapter 9) and address the matter of certification (in Chapter 10).

CHAPTER 8

SOME COMMENTS ON PART 6 OF IEC 65108

8.1 Overview

The technical content is contained in five informative annexes.

Annex A covers the general concept of Parts 2 and 3 and identifies the steps in their implementation. This functional capability has been covered in Chapter 2 and Appendix 1 of this book.

Annex B is an example of using the Part 6 tables for evaluating random hardware failure rates and probabilities. This is expanded in the following section and applied to the Chapter 11 exercise.

Annex C covers the mechanics of calculating the diagnostic coverage and safe failure fraction, which is covered in Chapter 3 and Appendix 4 of this book.

Annex D offers a methodology for determining the proportion of common cause failures. However, this book provides a more recent and calibrated model (BETAPLUS) in Chapter 6 and Appendix 3.

Annex E gives examples of the application of the safety-integrity tables given in Part 3. The tables given at the end of Chapter 4 of this book fulfil the same function. Furthermore, similar tables for hardware systematic failures were provided near the end of Chapter 3 of this book. IEC 61508 does not provide the latter.

8.2 The quantitative tables (Annex B)

This Annex provides:

- Recommended steps to follow for calculating the proposed safety system failure rate or PFD
- Detailed failure fate/PFD equations for a number of commonly used configurations
- A set of tables based on different proof-test intervals to enable quick estimates for a system failure rate/PFD to enable designers to make judgements on the viability of alternative proposed safety system configurations

The first two items are covered in Chapters 6 and 7 of this book. **Note that the IEC 61508 tables were based on the traditional Markov equations which have recently been revised by K.G.L. Simpson and published in the *Safety and Reliability Society Journal* (Summer 2002).**

USE OF THE PART 6 TABLES

In order to use these tables the overall safety system should be divided into sub-systems such that, in reliability modelling terms, these sub-systems are in series. Thus the individual sub-system failure rates (or PFDs) can be added together to obtain the overall system failure rate/PFD.

The majority of add-on safety systems consist of three major sub-systems (Sensors, Logic, Final elements). In some cases one or more of these major sub-systems can be further divided into smaller sub-systems.

The total failure rate/PFD for the safety system will be obtained from the sum of the sub-systems. The method of estimating each of the sub-system failure rates/PFDs is the same.

The failure rate/PFD for a sub-system can be estimated by stating a number of hardware reliability parameters that best describe the proposed components and their configuration. The failure rate/PFD is then obtained from the appropriate table.

There are a number of tables in Part 6 each catering for a specific proof-test interval:

- For **continuous systems** there are tables for proof-test intervals of 1, 3, 6 and 12 months, and the tables present failure rates (per hour).
- For **low demand systems** there are tables for proof-test intervals of ½, 1, 2 and 10 years, and the tables present PFDs (dimensionless).

Each table has columns representing component failure rates from 0.1×10^{-6} per hr to 50×10^{-6} per hr. These failure rates are the *total* component failure rates and include both the so-called 'safe' failures and so-called 'dangerous' failures. The tables assume that the ratio of 'safe' to 'dangerous' failures is 1:1.

For each of the failure rate columns there are three options for the common cause failure BETA factor as applied to failures not revealed by auto-test (2%, 10%, 20%). Different values (1%, 5%, 10%) are assumed for the auto-tested proportion of the failure rate.

Each table has five groups of rows, each group representing a common configuration (i.e. 1oo1 (simplex), 1oo2 (one out of two), 2oo2 (two out of two), 1oo2D (two out of two reverting to one out of one on detection of a faulty channel), and 2oo3 (two out of three)). For each of these rows there are four options for the diagnostic coverage (auto-test) of the components, namely, 0% (which represents less than 60%), 60% (which represents 60% to 90%), 90% (which represents 90% to 99%) and 99% (which represents greater than 99%).

Thus, for a range of component/sub-system configurations, it is possible to obtain a rough estimate of the suitability of a proposed safety system to meet a required SIL target. It should be kept in mind that each SIL covers an order of magnitude and thus the closeness of the actual component to the selected parameters in the tables has a fair degree of tolerance.

PFD for low demand mode of operation for a proof-test interval of 1 year (based on Table B3 of Part 6)

Configuration	*Diag. cov.*	λ = 1.0E–06			λ = 5.0E–06		
		β = 2%/1%	β = 10%/5%	β = 20%/10%	β = 2%/1%	β = 10%/5%	β = 20%/10%
1 out of 1	0%		$\mathbf{2.2 \times 10^{-3}}$			1.1×10^{-2}	
	60%		8.8×10^{-4}			4.4×10^{-3}	
	90%		2.2×10^{-4}			$\mathbf{1.1 \times 10^{-3}}$	
	99%		2.6×10^{-5}			1.3×10^{-4}	
1 out of 2	0%	5.0×10^{-5}	2.2×10^{-4}	4.4×10^{-4}	3.7×10^{-4}	1.2×10^{-3}	2.3×10^{-3}
	60%	1.9×10^{-5}	8.9×10^{-5}	1.8×10^{-4}	1.1×10^{-4}	4.6×10^{-4}	9.0×10^{-4}
	90%	4.5×10^{-6}	2.2×10^{-5}	4.4×10^{-5}	2.4×10^{-5}	1.1×10^{-4}	2.2×10^{-4}
	99%	4.8×10^{-7}	2.4×10^{-6}	4.8×10^{-6}	2.4×10^{-6}	1.2×10^{-5}	2.4×10^{-5}

Configuration	*Diag. cov.*	λ = 10.0E–06		
		β = 2%/1%	β = 10%/5%	β = 20%/10%
1 out of 1	0%		$\mathbf{2.2 \times 10^{-2}}$	
	60%		8.8×10^{-3}	
	90%		2.2×10^{-3}	
	99%		2.6×10^{-4}	
1 out of 2	0%	1.1×10^{-3}	2.7×10^{-3}	4.8×10^{-3}
	60%	2.8×10^{-4}	9.7×10^{-4}	1.8×10^{-3}
	90%	5.1×10^{-5}	2.3×10^{-4}	4.5×10^{-4}
	99%	4.9×10^{-6}	2.4×10^{-5}	4.8×10^{-5}

Note: 'Diag. cov.' refers to the diagnostic coverage (not the safe fail fraction). β = 2%/1% signifies 2% for failures revealed by normal proof test and 1% for those revealed by auto-test.

PFD for low demand mode of operation for a proof-test interval of 6 months (based on Table B2 of Part 6)

Configuration	*Diag. cov.*	λ = 1.0E–06			λ = 5.0E–06		
		β = 2%/1%	β = 10%/5%	β = 20%/10%	β = 2%/1%	β = 10%/5%	β = 20%/10%
1 out of 1	0%		1.1×10^{-3}			5.5×10^{-3}	
	60%		4.4×10^{-4}			2.2×10^{-3}	
	90%		1.1×10^{-4}			$\mathbf{5.7 \times 10^{-4}}$	
	99%		1.5×10^{-5}			7.5×10^{-5}	
1 out of 2	0%	2.4×10^{-5}	$\mathbf{1.1 \times 10^{-4}}$	2.2×10^{-4}	1.5×10^{-4}	5.8×10^{-4}	1.1×10^{-3}
	60%	9.1×10^{-6}	4.4×10^{-5}	8.8×10^{-5}	5.0×10^{-5}	2.3×10^{-4}	4.5×10^{-4}
	90%	2.3×10^{-6}	1.1×10^{-5}	2.2×10^{-5}	1.2×10^{-5}	5.6×10^{-5}	1.1×10^{-4}
	99%	2.6×10^{-7}	1.3×10^{-6}	2.6×10^{-6}	1.3×10^{-6}	6.5×10^{-6}	1.3×10^{-5}

Configuration	*Diag. cov.*	λ = 10.0E–06		
		β = 2%/1%	β = 10%/5%	β = 20%/10%
1 out of 1	0%		$\mathbf{1.1 \times 10^{-2}}$	
	60%		4.4×10^{-3}	
	90%		1.1×10^{-3}	
	99%		1.5×10^{-4}	
1 out of 2	0%	3.7×10^{-4}	1.2×10^{-3}	2.3×10^{-3}
	60%	1.1×10^{-4}	4.6×10^{-4}	9.0×10^{-4}
	90%	2.4×10^{-5}	1.1×10^{-4}	2.2×10^{-4}
	99%	2.6×10^{-6}	1.3×10^{-5}	2.6×10^{-5}

EXAMPLE IN THE USE OF THE TABLES

The following example is based on the Chapter 11 exercise.

Sensor sub-system – Pressure transmitter

Property	*Actual*	*Nearest fit used in table*
Total failure rate (per hr)	2×10^{-6}	1×10^{-6}
Diagnostic cover (%)	0	0
CCF (β %)	N/A	N/A
Proof-test period	1 year	1 year
Configuration	1oo1	1oo1

From Table B3 of Part 6 using the above parameters yields a result of PFD = $\mathbf{2.2 \times 10^{-3}}$

Logic sub-system – PES

Property	*Actual*	*Nearest fit used in table*
Total failure rate (per hr)	5×10^{-6}	5×10^{-6}
Diagnostic cover (%)	90	90
CCF (β %)	N/A	N/A
Proof-test period	1 year	1 year
Configuration	1oo1	1oo1

From Table B3 of Part 6 using the above parameters yields a result of PFD = $\mathbf{1.1 \times 10^{-3}}$

Final element – Ball valves

Property	*Actual*	*Nearest fit used in table*
Total failure rate (per hr)	8×10^{-6}	10×10^{-6}
Diagnostic cover (%)	0	0
CCF (β %)	N/A	N/A
Proof-test period	1 year	1 year
Configuration	1oo1	1oo1

From Table B3 of Part 6 using the above parameters would yield a result of PFD = $\mathbf{2.2 \times 10^{-2}}$ for each valve

However, it is here that we need to take careful note of the underlying assumptions explained at the beginning of this chapter. It was pointed out that the tables assume a 50% fail dangerous mode. In this example the sprung to close valve has 10% dangerous failures as a result of which the tables have given a very pessimistic result. However, had the sprung to close failure mode been 90% the reverse would have been the case and an optimistic result obtained. Thus great care must be taken in their use.

As a result we shall divide the answer by 5 (i.e. the ratio of 10% to 50%). This yields 4.4×10^{-3} per valve.

Thus the total PFD $= 2.2 \times 10^{-3} + 1.1 \times 10^{-3} + 2 \times (4.4 \times 10^{-3}) = \mathbf{1.2 \times 10^{-2}}$

Recalculating the above for the proposed modifications in Chapter 11 involves Table B2.

The reader may wish to check the final calculation which becomes:

Total PFD $= 1.1 \times 10^{-4} + 5.7 \times 10^{-4} + 2 \times (1.1 \times 10^{-2})/5 = \mathbf{5.1 \times 10^{-3}}$

8.3 The software safety-integrity tables (Annex E)

The Standard provides two examples (one at SIL 2 and one at SIL 3) whereby the Part 3 tables for the SIL in question are reproduced containing a sample interpretation for each and every item.

This necessarily leads to a somewhat fragmented and lengthy presentation of an assessment, making it difficult to see the overall picture. Chapters 3, 4 and 5 of this book have attempted to provide a more concise format without the need to repeat requirements common to activities throughout the life-cycle.

Chapter 9

Second tier and related guidance documents

Some of the following documents are referred to as 'second tier' guidance in relation to IEC 61508. Due to the open-ended nature of the statements made, and to ambiguity of interpretation, it cannot be said that conformance with any one of them automatically infers compliance with IEC 61508.

However, they cover much the same ground as each other albeit using slightly different terms to describe documents and life-cycle activities.

Figure 9.1 illustrates the relationship of the documents to IEC 61508. A dotted line indicates that the document addresses similar issues whilst not strictly being viewed as second tier.

9.1 IEC International Standard 61511: Functional safety – safety instrumented systems for the process industry sector

IEC 61511 is intended as the process industry sector implementation of IEC 61508.

It gives application specific guidance on the use of standard products for use in 'safety instrumented' systems using the proven-in-use justification. The guidance allows the use of field devices to be selected based on proven-in-use for application up to SIL 3 and for standard off-the-shelf PLCs for applications up to SIL 2.

Since it is the first sector specific IEC document to follow publication of 61508, it is considered sufficiently important to have been given a separate chapter (Chapter 5) in this second edition.

9.2 Institution of Gas Engineers and Managers IGEM/SR/15: Programmable equipment in safety-related applications – 3rd Edition and Amendment

This is the Gas Industry second tier guidance to IEC 61508. It is suitable for oil and gas and process applications.

SR/15 describes both quantitative and risk matrix approaches to establishing target SILs but a preference for the quantitative approach is stressed. More specific design guidance is given for pressure and flow control, gas holder control, burner control and process shutdown systems.

An amendment, published in 2002, addresses the setting of maximum tolerable risk targets (fatality levels), cost per life saved and also includes a checklist schedule to aid conformity in the rigour of carrying out assessments. The tolerable risk targets were shown in Chapter 2 of this book. The checklist for rigour of assessment covers the items listed in Appendix 2 of this book. The term 'Required' is used to replace the more cumbersome 'Highly Recommended' of IEC 61508. The document has 107 pages (16 for the addendum). The fourth edition will be published in 2004.

9.3 UKOOA: Guidelines for Process Control and Safety Systems on Offshore Installations

Currently at Issue 2 (1999), this United Kingdom Offshore Operators Association guide offers guidance for control and safety systems offshore. The sections cover:

- The role of control systems in hazard management
- Categorisation of systems (by hazard and application)
- System design
- Equipment design
- Operation and maintenance

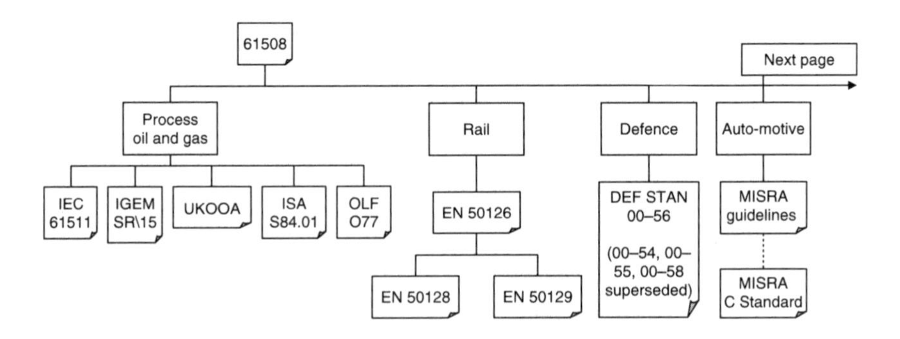
61508
Next page
Process oil and gas
Rail
Defence
Auto-motive
IEC 61511
IGEM SR\15
UKOOA
ISA S84.01
OLF O77
EN 50126
EN 50128
EN 50129
DEF STAN 00–56
(00–54, 00–55, 00–58 superseded)
MISRA guidelines
MISRA C Standard

Figure 9.1

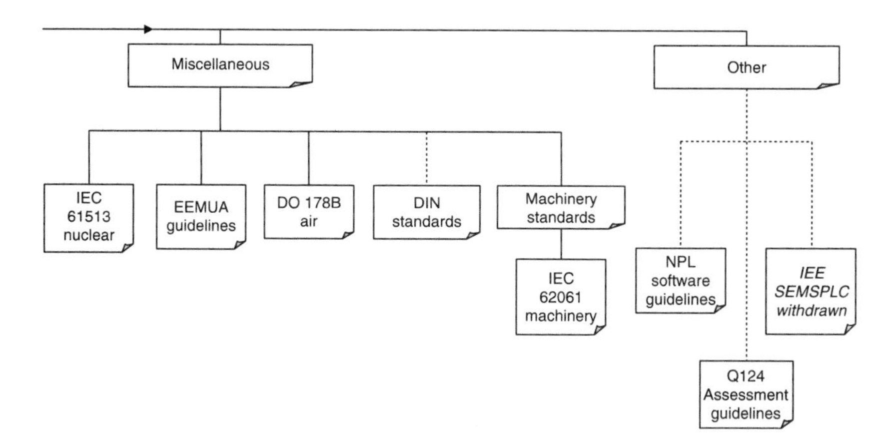
Miscellaneous
Other
IEC 61513 nuclear
EEMUA guidelines
DO 178B air
DIN standards
Machinery standards
IEC 62061 machinery
NPL software guidelines
IEE SEMSPLC withdrawn
Q124 Assessment guidelines

There is an Appendix addressing software in safety-related systems. Safety-integrity levels are described in a similar way to IEC 61508. However, SIL 4 is not recommended.

The setting of SIL targets for protection systems is initially approached by a risk graph method (reproduced in Chapter 3 by kind permission of UKOOA) rather than using a quantitative approach. Nevertheless, quantitative assessment of SIL targets is referred to as an alternative to the risk graph.

Unlike IEC 61508, risk graphs are also offered for both environmental and for loss of production. Management systems and competence are called for as they are in IEC 61508. It has 55 pages.

9.4 Instrumentation Systems and Automation Society S84.01, 1996: Application of Safety Instrumented Systems for the Process Industries

The Instrumentation Systems and Automation Society (USA) is an International Society for Measurement and Control. They developed S84 as a response to IEC 61508 and it is intended as applications specific second tier guidance. It adopts the E/E/PES mnemonic in respect of safety instrumented systems (SIS) namely the sensors, logic solving and final elements in much the same way as IEC 61511.

Only three SILs are defined, being equivalent to SILs 1–3 of IEC 61508. Integrity greater than a PFD of 10^{-4} is not acknowledged, thereby implying the need for more levels of protection to achieve high integrity. The SILs are applicable to the SIS rather than to external risk reduction measures.

A risk matrix, involving consequence severity, likelihood of occurrence and effectiveness of protection, is provided for obtaining SIL targets. Simple architecture diagrams suggest the SILs likely to be achieved by given amounts of redundancy.

Again, a life-cycle approach is adopted from process design, through procurement and installation and including operations, maintenance, modifications and decommissioning. The process starts with a Safety Requirements Specification and moves through the life-cycle with requirements similar to IEC 61508.

An Annex provides detailed design guidance on issues such as sensor diversity, communications, embedded software and electromechanical devices. For example, the guidance on sensor diversity suggests:

- SIL 1: Single sensor likely to be suitable
- SIL 2: Redundancy (identical) with separation
- SIL 3: Redundancy (diverse) with separation

The entire document is 107 pages and adopts a similar approach to IEC 61508, although there is some emphasis on the avoidance of systematic faults. It is written in a straightforward manner.

9.5 Recommended Guidelines for the Application of IEC 61508 and 61511 in the Petroleum Activities on the Norwegian Continental Shelf OLF-077

Published by the Norwegian Oil Industry Association, this 46 page document provides typical safety loops along with the recommended configuration and anticipated SIL. It should be noted that these recommended SILs are typically ONE LEVEL higher than would be expected from the conventional QRA approach described in Chapter 2 of this book.

This is the result of a Norwegian law which states that any new standard, associated with safety, must IMPROVE on what is currently being achieved. Therefore the authors of OLF-077 assessed the current practices in the Norwegian sector and calculated the expected PFDs for each safety loop and determined which SIL band they fitted.

It should also be noted that the guidelines give failure rate figures for systematic as well as random hardware failures.

A typical example of a recommended loop SIL is shown in Figure 9.2.

9.6 European Standard EN 50126: Railway Applications – The Specification and Demonstration of Dependability, Reliability, Maintainability and Safety (RAMS)

The development of standards and standard approaches for the design and demonstration of the safety of (in the main) programmable electronic systems for railway-related application has led to the development of a suite of standards. This suite

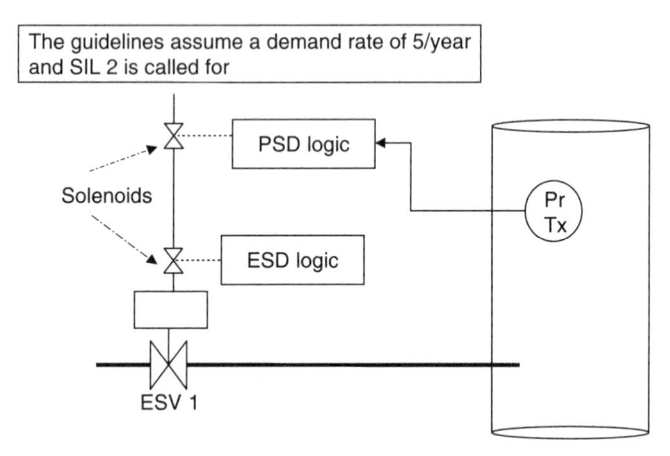

Figure 9.2
OLF-077 – process shutdown functions: PAHH, LAHH, LALL PAHH function

provides both an approach that supports the (general) requirements of IEC 61508, and also a means to encourage European rail industry interoperability. This latter element has become increasingly important through the development of Technical Specifications for Interoperability (TSIs) for railway lines classified as suitable for High Speed and Conventional operation. The certification of European railway equipment and systems as 'fit for purpose' requires a certification of their 'interoperability', that is, their ability to be applied to any member state railway, primarily in order to encourage competition and sustainable growth within the EU member states' railway undertakings.

EN 50126 is effectively the European-wide Rail Industry second tier general guidance (1999) for IEC 61508. It is often referred to as 'the RAMS standard', as it addresses both reliability and safety issues. EN 50126 is intended to cover the railway system in total, while the companion standards, EN 50128 and EN 50129 are more specific. CENELEC describe standard 50126 as being 'intended to provide Railway Authorities and the railway support industry throughout the European Community with a process which will enable the implementation of a consistent approach to the management of RAMS'.

Risks are assessed by the 'risk classification' approach (described at the end of Chapter 2) whereby severity, frequency, consequence etc. are specified by guidewords and an overall 'risk classification matrix' obtained. 'Intolerable', 'ALARP' and 'Negligible' categories are thus derived and one proceeds according to the category assessed. The acceptance (or otherwise) of risk is based on choosing a risk acceptance (or hazard tolerability) scheme, the principles of which can be applied throughout the member states (or indeed by other railway authorities). Examples of acceptable risk classification schemes given include 'ALARP' in Great Britain 'GAMAB' (Globalement au moins aussi bon) in France and 'MEM' (Minimum Endogenous Mortality) in Germany. In general terms, the first two schemes deal with global or total risk, whereas the scheme applied in Germany assesses risk to individuals.

The Standard is life-cycle based, using the 'v-curve' life-cycle approach. This means that requirements are stated (and subsequently verified and validated) throughout the concept, specification, design and implementation stages of a project. Input and outputs (i.e. deliverables) are described for the life-cycle activities.

EN 50126 is concerned with the more general specification for the RAMS requirements of a total railway system and the necessary risk assessment, including development of SIL targets and their subsequent satisfactory demonstration, which includes the control of the activities. CENELEC (draft) Standard pr**EN 50128**, 2002 **'Railway Applications: Software for Railway Control and Protection Systems'** covers the requirements for software for railway control and protection systems. It is stated by CENELEC that 'The standard specifies procedures and technical requirements for the development of programmable electronic systems for use in railway control and protection applications. The key concept of the standard is the assignment of levels of integrity to software. Techniques and measures for 5 levels of software integrity are detailed.'

BS EN 50129, 2002 **'Railway Applications, Safety-related Electronics for Signalling'** provides details for (hardware and software) for railway control and protection systems. EN 50129 has been produced as a European standardisation document defining requirements for the acceptance and approval of safety-related electronic systems in the railway signalling field. The requirements for safety-related hardware and for the overall system are defined in this Standard. It is primarily intended to apply to 'fail-safe' and 'high integrity' systems such as main line signalling.

The requirements for 50128 and 50129 are those that are most similar (in detail) to the requirements of IEC 61508. Thus the suite of three Standards provides the overall response to IEC 61508, with the three railway specific documents being roughly equivalent to the Part 1, 2, 3 structure of IEC 61508.

9.7 UK MOD documents

The Ministry of Defence has had its own suite of standards covering much the same ground. Currently 00–56 is being rewritten (as Issue 3.0) to supersede the earlier suite which is nevertheless summarised below.

(a1) Defence Standard 00–56 (Issue 2.0): Hazard analysis and safety classification of the computer and programmable electronic system elements of defence equipment

This was akin to Part 1 of IEC 61508. Whereas 00–55 (see below) addressed software this earlier 00–56 encompassed the entire 'safety critical' system. It called for HAZOPS (HAZard and OPerability Studies) to be carried out on systems and subsystems of safety-related equipment supplied to the UK MOD. There were tables to assist in the classification and interpretation of risk classes and activities are called for according to their severity. This is a risk graph approach which establishes SIL targets. Responsibility for safety was to be formally defined as well as the management arrangements for its implementation. It was intended that 00–56 harmonised with RTCA DO-178B/(EUROCAE ED-12B) (see 9.12 below) and that it should be compatible with IEC 61508. It was in two parts.

(a2) Defence Standard 00–56 (DRAFT Issue 3.0): Safety management requirements for defence systems

The proposed Issue 3 will supersede Issue 2.0 and the other documents listed below. It contains less in the way of techniques which might be published, later, as a handbook.

The new structure will be:

Part 1: Requirements: This is largely an exhortation to establish safety management, identify hazards and establish a safety case which will reflect risk assessments and the subsequent demonstration of tolerable risks following appropriate risk reduction.

Part 2: Code of Practice: provides more detail on the practices to be adopted to satisfy Part 1. It will comprise four volumes:

1. Interpretation of Part 1 – Somewhat repetitive, although with more detail such as items for the content of safety cases, aspects of hazard identification etc.
2. Risk management – Addressing issues such as HAZID, risk classification and SILs (previously covered in 00–56 Issue 2.0)
3. Software – A successor to 00–55 (below)
4. Electronic hardware – A successor to 00–54 (below)

(b) (*To be superseded*) Defence Standard 00–54: Requirements for safety-related electronic hardware in defence equipment

This complements 00–55 and 00–56 by covering the hardware aspects. It is life-cycle based and covers much the same ground as IEC 61508 Part 2. The guidance is tailored in rigour according to the SIL. There are 65 pages in total.

(c) (*To be superseded*) Defence Standard 0055: The procurement of safety critical software in defence equipment

This is akin to Part 3 of IEC 61508 and has superseded the old MOD 00–16 guide to achievement of quality in software. It is far more stringent and is perhaps one of the most demanding standards in this area.

Whereas the majority of the documents described here are for guidance, 00–55 is a standard and is intended to be mandatory on suppliers of 'safety-critical' software to the MOD. It is unlikely that the majority of suppliers are capable of responding to all of its requirements but the intention is that, over a period of time, industry evolves to adopt it in full.

It deals with software rather than the whole system and its major requirements include:

- The non-use of assembler language
- The use of static analysis
- A preference for formal methods

- The use and approval of a safety plan
- The use of a software quality plan
- The use of a validation plan
- An independent safety auditor

There are 75 pages in the two parts.

(d) (*To be superseded*) Standard 00–58: A guideline for HAZOP studies on systems which include programmable electronic systems

As the title suggests this Standard describes the HAZOP process in the context of identifying potentially hazardous variations from the design intent. Part 1 is the requirements and Part 2 provides more detailed guidance on such items as HAZOP guide words for particular types of system, team roles, recording the study etc. There are 86 pages in the two parts.

9.8 MISRA (Motor Industry Research Assoc) 1994: Development guidelines for vehicle based software

The SIL categorisation (0 to 4) is qualitatively defined as follows and is to be applied as a result of the failure mode that attracts the highest SIL. The document was developed in 1994 and the SILs are qualitatively defined rather than mapping to the numerical ranges shown in Table 1.1 of Chapter 1. Again, the guidance is based on the software life-cycle. It has 82 pages.

Controllability	*Acceptable failure rate*	*SIL*
Uncontrollable	Extremely improbable	4
Difficult to control	Very remote	3
Debilitating	Remote	2
Distracting	Unlikely	1
Nuisance only	Reasonably possible	0

9.9 The MISRA C Coding Standard

The document (currently being revised) provides a sub-set of the C language for use up to SIL 3. It contains many rules for the use of the language in safety-related applications.

It starts with the premise that the full C should not be used for safety-related systems. It explains the need for a sub-set and describes how to use it but, nevertheless, assumes familiarity and competence with the language. It recommends against the use of assembly language in this context.

The contents can be summarised as:

1. **Background:** covering language insecurities, compiler issues, safety-related uses and standardisation
2. **Vision:** a chapter on the rationale for the sub-set
3. **Developing the sub-set**
4. **Scope:** covering language issues, applicability, SILs (C++ is excluded) and auto-code
5. **Using MISRA C:** a chapter on managing and implementing the sub-set
6. **Introduction to the rules:** a general introduction
7. **Rules:** the detail including character sets, initialisation, control flow, pointers, libraries etc.

Further information can be obtained from www.misra.org.uk.

9.10 IEC International Standard 61513: Nuclear Power Plants – Instrumentation and control for systems important to safety – general requirements for systems

Many of the existing standards, which were applicable to the nuclear sector prior to the emergence of IEC 61508, generally adopted a similar approach to the draft 61508. These existing standards are either from IEC or IAEA. Thus the nuclear sector standard IEC 61513 primarily links these existing standards to IEC 61508. The IEC existing standards are **60880, 60987, 61226 and 60964**, and the existing IAEA standards are primarily **NS-R-1, 50-SG-D1, 50-SG-D3 and 50-SG-D8.**

These standards present a similar overall safety cycle and system life-cycle approach as in IEC 61508 with more in-depth details at each stage compared to IEC 61508. **IEC 60964** covers the identification of the required safety function applicable to power plants and **IEC 61226** provides system categorisation for different types of safety functions. The SIS design is then covered by **IEC 60987** for hardware design and **IEC 60880** for software design. **IAEA 50-C-D, now NS-R-1**, covers the overall Safety Design, **50-SG-D1** gives the Classification of Safety Functions, **50-SG-D3** covers all Protection Systems and **50-SG-D8** provides the requirements for the Instrumentation and Control Systems.

The current standards do not directly use the SAFETY INTEGRITY LEVELS as given in IEC 61508. The standards use the existing categorisation (IEC 61226) A, B or C. These are related to 'Safety Functions', A = highest and C = lowest. IEC 61513 adds corresponding system classes, 1 = highest and 3 = lowest, where:

Class 1 system can be used for Cat A, B or C
Class 2 system can be used for Cat B or C
Class 3 system can be used for Cat C

Categorisation A is for safety functions, which play a principal roll in maintenance of NPP safety
Categorisation B is for safety functions that provide a complementary role to category A
Categorisation C is for safety functions that have an indirect role in maintenance of NPP safety

No specific reliability/availability targets are set against each of these categories or classes. There is, however, a maximum limit set for software-based systems of 10^{-4} PFD. More generally the reliability/availability targets are set in the Plant Safety Design Base and can be set either quantitatively or qualitatively. There is a preference for quantitative plus basic requirements on layers and types of protection.

Class 1/Categorisation A is generally accepted as being equivalent to SIL 3
Class 2/Categorisation B is generally accepted as being equivalent to SIL 2
Class 3/Categorisation C is generally accepted as being equivalent to SIL 1

ARCHITECTURAL CONSTRAINTS do not have a direct relationship with the tables in IEC 61508 Part 2, but are summarised below:

CAT A Shall have redundancy, to be fault tolerant to one failure, with separation. Levels of self-test are also given.

CAT B Redundancy is preferred but a simplex system with adequate reliability is acceptable, again levels of self-test are given.

CAT C Redundancy not required. Reliability needs to be adequate, self-test required.

GENERAL DESIGN REQUIREMENTS

Within this Standard and the related Standard there is significantly more guidance on each of the steps in the design. In particular:

- Human factors
- Defences against common cause failures
- Separation/Segregation
- Diversity

There are mapping tables for relating its clauses to the clause numbers in IEC 61508.

9.11 EEMUA Guidelines, Publication No. 160: Safety related instrument systems for the process industry (including programmable electronic systems)

These were published, in 1989, by EEMUA (Engineering Equipment and Materials Users Association) in response to the HSE PES guidance mentioned in Chapter 1. They were produced well before the emergence of IEC 61508 drafts and are not thought to be widely used these days.

The document defines four categories of equipment known as 0, 1, 2 and 3. Category 0 is the highest in the sense of self-acting protective devices (e.g. relief valves). Category 1 is for protective systems which require external energy (e.g. relay and electronic systems). Category 2 is for systems protecting the environment and Category 3 for systems protecting production. Requirements are provided against each category and the applicability (acceptability or otherwise) of 'self-acting', 'non-programmable', 'fixed program', 'limited variability' and 'full variability' systems is given for each. Some of the guidance is specific to architectures such as 'two out of three', 'one out of two' etc. The document has 82 pages.

9.12 RTCA DO-178B/(EUROCAE ED-12B): Software considerations in airborne systems and equipment certification

This is a very detailed and thorough Standard which is used in civil avionics to provide a basis for certifying software used in aircraft. Drafted by a EUROCAE/RTCA committee, DO-178B was published in 1992 and replaces an earlier version published in 1985. The qualification of software tools, diverse software, formal methods and user-modified software are now included.

It defines five levels of software criticality from A (software which can lead to catastrophic failure) to E (no effect). The Standard provides guidance which applies to levels A to D.

The detailed listing of techniques covers:

- SYSTEMS ASPECTS: including the criticality levels, architecture considerations, user modifiable software.
- THE SOFTWARE LIFE-CYCLE
- SOFTWARE PLANNING
- DEVELOPMENT: including requirements, design, coding and integration.
- VERIFICATION: including reviews, test and test environments.
- CONFIGURATION MANAGEMENT: including baselines, traceability, changes, archive and retrieval.
- SOFTWARE QUALITY
- CERTIFICATION
- LIFE-CYCLE DATA: describes the data requirements at the various stages in the life-cycle.

Each of the software quality processes/techniques described in the Standard is then listed (10 pages) and the degree to which they are required is indicated for each of the criticality levels A to D. The document has 67 pages.

9.13 DIN V Standards

The following two German standards would, prior to IEC 61508, have been used for product certification. However, they would now be applied in conjunction with IEC 61508.

(a) DIN V 19250: Measurement and control, fundamental safety aspects for measuring and control protective equipment

Although dated 1989, this document describes the IEC 61508 concept of successive risk reduction by one or more protection measures. Risk classification, together with a risk graph (almost identical in structure to Figure 2.2), leads to eight categories of safety integrity. In Chapter 10 we show how these map to the 4 SILs in IEC 61508.

System features (e.g. drift failure, fault accumulation, accidental fault) are defined and technical and non-technical protection measures are described for each. The approach is failure based and does not address the familiar safety life-cycle of more recent standards. The document has 44 pages.

(b) VDE 0801: Principles for computers in safety-related systems

This is life-cycle based and deals with methods of error avoidance in the development of both hardware and software. It invokes the risk categories of DIN V 19250 above. There are comprehensive technical checklists and the document is 172 pages long.

9.14 Documents related to machinery

Three Standards are mentioned here. Items (a) and (b) are harmonised under the EC Machinery Directive. Item (c) is the sector specific draft IEC document IEC 62061.

(a) EN 954-1 Safety of machinery: Safety-related parts of control systems

This document was developed primarily for non-programmable (mechanical and hydraulic) systems. Rather than being a system-based life-cycle approach, it is concerned with the safety-related parts of control systems and addresses fault tolerance. There are five categories described (B, 1, 2, 3 and 4) for machinery control systems and a risk graph, not unlike Figure 2.3, addresses severity of injury, frequency of exposure and possibilities for avoidance.

However, despite the apparent similarity with the IEC 61508 SILs, the above categories are not held to be equivalent and, indeed, are not even hierarchical in their '1–4' sequence. Nevertheless it is difficult not to draw a comparison.

In Category B (the lowest) a single fault can be permitted to lead to loss of the safety function. Reliability in use and test data are called for as evidence of conformity.

In Category 1 well-tried and proven components are called for and life testing and failure mode definition are mentioned.

In Category 2 start-up checks (manual or automatic) are required to prove the safety function.

In Category 3 a single fault may not lead to loss of the safety function and in Category 4 diversity is called for.

In Category 4 there must be 'no fail to danger state'. This implies a probability of failure on demand of ZERO which is clearly impossible.

The reader must make his own interpretation of the foregoing.

The Standard has been much used and it remains to be seen if the emerging IEC 62061 will actually supersede EN 954-1.

(b) EN60204-1 Safety of machinery: Electrical equipment of machines

This is somewhat of a hybrid, covering both electrical and control equipment.

(c) IEC International Standard 62061: Safety of machinery – functional safety of electronic and programmable electronic control systems for machinery

The scope is evident from the title and compliance is consistent with the requirements of IEC 61508. The usual life-cycle activities and documents are described. It remains to be seen whether this will supersede EN 954-1.

9.15 Validation of measurement software, NPL

This is currently a draft document (2002), published by the National Physical Laboratory. It has two main sections: Part 1 Management and Part 2 Technical.

This document defines 'Measurement Software Levels' (0–4). The issue of the match between the Measurement Software Level and the 61508 SIL has yet to be clarified although in general (it is claimed) a 'Measurement Software Level' of x is required to achieve a SIL of x.

Chapter 5 of Part 1 defines the targeting of the five 'Levels'. This consists of a qualitative description of the application for which each 'Level' is applicable rather than the quantified risk approach of 61508. In summary they are:

- Level 0: Very simple software with no problems revealed in the analysis.
- Level 1: 'Simple' data processing.
- Level 2: At least one major unquantifiable aspect of the software.
- Level 3: Either complex software (difficult to validate) or software with significant problems.
- Level 4: Involves the highest measurement integrity.

The NPL guide gives levels for the measurement software whilst 61508 refers to the whole system. It then describes the quantified IEC 61508 SILs but does not offer any mapping between them and the NPL 'Levels'. IEC 61508 involves a SIL target for the overall system, which automatically becomes a requirement for each of the sub-systems and components. The NPL document, on the other hand, encourages different (but suitable) targets for various parts of a system, depending on the measurement integrity requirements of each part.

The various requirements of the Measurement Software Levels are approximately the same as for the 61508 SIL levels. Table 11.2 of the guidelines outlines the items applicable to each 'Level'. The requirements in section 12 of the guidelines are approximately the same as those for the corresponding 61508 SILs.

However, the ambiguity of interpretation of such terms as 'inspection', 'review', 'static analysis', 'verification' throughout the literature makes it impossible to establish a formal equivalence between the requirements of the two documents.

9.16 IEE Publication, SEMSPLC, 1996: Safety-related application software for programmable logic controllers

This 170 page document was an interpretation, at the time, of the draft 61508 Standard. It provides guidance specific to programmable logic controllers.

All the life-cycle phases are addressed and design guidance is offered for PLC specific items covering the same headings as the software aspects of the Standard.

The document has recently been withdrawn. Information is available on the IEE website.

9.17 Technis Guidelines, Q124, 2004: Demonstration of Product/System Compliance with IEC 61508

This 32 page document provides a framework for demonstration/ certification of either products or systems (be that by self demonstration, third party assessment or certifying body). It is intended for use by experienced functional safety professionals and offers a realistic level of rigour whilst allowing assessors scope for interpretation. It is available from Technis (see end of this book).

CHAPTER 10

DEMONSTRATING AND CERTIFYING CONFORMANCE

10.1 Demonstrating conformance

One might wish to demonstrate (or even certify) conformance to the requirements of IEC 61508 in two respects.

FIRST: That an organisation can demonstrate the generic capability to produce such a product or system.

SECOND: That a product or system design meets the requirements outlined in the preceding chapters.

In the first case it is the raft of procedures and work practices, together with the competence of individuals, which is being assessed. This is known as the FUNCTIONAL SAFETY CAPABILITY (FSC) of an organisation and is demonstrated by its quality management system and evidenced by documented audits and examples of the procedures being used.

In the second it is the design, and the life-cycle activities, of a particular product which is being assessed. This is demonstrated by specifications, design documents, reviews, test specifications and results, failure rate predictions, FMEAs to determine safe failure fraction and so on.

In practice, however, it is not really credible to audit one without evidence of the other. FSC needs to be evidenced by at least one example of a product or project and a product's conformance needs to be evidenced by documentation and life-cycle activities which show overall capability.

In fact there are currently five subdivisions (rather than the two described above) and these are explained in the next section.

10.2 The current framework for certification

Most people in industry are, by now, well aware of the certification framework for ISO 9000. UKAS (United Kingdom Accreditation Service) **accredits** organisations to be able to **certify** clients to ISO 9001-2000.

There are over 100 accredited bodies (in the UK alone) offering ISO 9000 certification and many thousands of organisations who have been certified, by them, to the ISO 9000 Standard. There is only one aspect of certification – one either meets the Standard or one does not.

The situation for IEC 61508 is rather different and less well developed.

First, there are the two aspects to the certification (namely the organisation or the product). In the case of the product, unlike 9000, there are four levels of rigour against which to be certified (SILs 1–4).

Following a DTI initiative in 1998/9, a framework was developed by CASS Ltd (Conformity Assessment of Safety-related Systems). One motive for this was to erode differences in approach across application sectors and thereby improve the marketability of UK safety-related technology. Another was to prevent multiple assessments and also to meet the need for the ever-increasing demand for assessment of safety-related equipment. The CASS framework suggests FIVE types of assessment:

Type 5: Functional Safety Capability Assessment (known as FSCA)
  Described in Chapter 2 and catered for by Appendix 1 of this book

Type 4: Safety Requirements Assessment
  Addressing those who carry out the risk assessments

Type 3: Operations and Maintenance Assessment
  For those carrying out operations and maintenance

Type 2: Application Specific Systems Assessment
  This is the overall assessment of SIL targets and of whether a system meets them, as addressed throughout this book

Type 1: Non-Application Specific Component Assessment
This meets the need for an SIL requirement on a component whose SIL is dependent on the application. In these circumstances one would only be able to address a sub-set of the total requirements, e.g. (Part 3 of IEC 61508). Note the comments in Section 2.2.2 concerning component 'SILs'.

At present UKAS (United Kingdom Accreditation Service) have accredited one body (SIRA Certification Service) to certify to the Type 5 conformance. Currently eight organisations (including the co-author's company Silvertech Ltd) have the Type 5 Functional Safety Certification. Figure 10.1 shows the current framework. UKAS have also accredited BASEEFA 2001 to certify Type 2 systems for Part 2 of the Standard. SIRA is also moving towards a total Type 1 and 2 scheme.

At the time of writing CASS has published FSCA schedules (Type 5). Furthermore, since assessor competence is important, CASS has interviewed and approved a small number of assessors.

There is a strong demand from industry for Type 1 and Type 2 certification particularly since the Type 2 systems designers require demonstrations of 61508 SIL conformance from their Type 1 component suppliers.

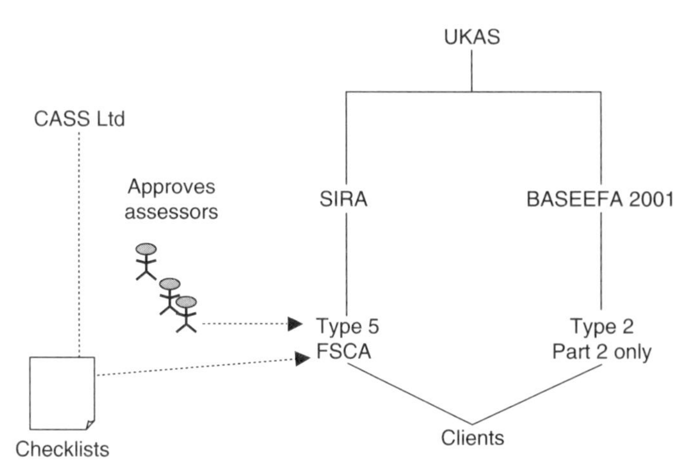

Figure 10.1
Certification framework

10.3 Self-certification (with independent assessment)

There is nothing to prevent self-assessment, either of one's Functional Safety Capability, as an organisation, or of the Safety Integrity Level of a product or design. Indeed this can be, and often is, as rigorous as the SIRA certification process. In any case for Type 1 and Type 2 demonstrations there is no other really practical route in the UK (see Section 10.4).

Third party involvement in the assessment, whilst not essential, is desirable to demonstrate impartiality and one requires a safety professional specialising in this field. The Safety and Reliability Society is associated with the Engineering Council, maintains appropriate standards for admission to corporate membership and membership would be one factor in suggesting suitability. Suitable consultants should have dealt with many other clients and have a track record concerning IEC 61508. Examples would be papers, lectures, assessments and contributions to the drafting of the Standard. This would serve to demonstrate that some assessment benchmark has been applied.

As a minimum self-assessment requires the following.

10.3.1 Functional Safety Capability as part of the Quality Management System

This is described in Chapter 2, being one of the requirements of Part 1 of IEC 61508. Appendix 1 of this book provides a template procedure which would be integrated into an organisation's quality management system.

The organisation would show evidence of a succession of audits and reviews of the procedure in order to claim compliance. Compliance with ISO 9001 is strongly indicated if one is aiming to claim functional safety compliance. The life-cycle activities are so close to the ISO 9001 requirements that it is hard to imagine a claim which does not include them. The ISO 9001 quality management system would need to be enhanced to include:

- Safety-related competencies (see Section 2.1.2)
- Functional safety activities (Appendix 1)
- Techniques for (and examples of) assessment (Chapters 6 and 7)

The scope of the capability should also be carefully defined because no one organisation is likely to be claiming to perform every activity described in the life-cycle. Examples of scope might include:

- Design and build of safety-related systems (Type 2)
- Design and build of safety-related instrumentation (Type 1)
- Assessment of SIL targets and of compliance of systems (Type 4)
- Maintenance of safety-related equipment (Type 3)

10.3.2 Application of 61508 to projects/products

In addition to the procedural capability described in Section 10.3.1 a self-assessment will also need to demonstrate the completion of at least one project together with a safety-integrity study.

The tables at the end of Chapters 3 and 4 of this book provide a means of formally recording the reviews and assessments. Chapters 11 and 12 show examples of how the quantitative assessments can be demonstrated.

10.3.3 Rigour of assessment

In addition to the technical detail suggested by Section 10.3.2 above, there needs to be visible evidence that sufficient aspects of assessment have been addressed. The 'assessment schedule' checklist in Appendix 2 of this book provides a formal checklist which allows one to demonstrate the thoroughness of an assessment.

It has to be acknowledged that third party assessment does involve additional cost for perhaps little significant added value in terms of actual safety integrity. Provided that the self-assessments are conducted under a formal quality management system, with appropriate audits, and provided also that competency of the assessors in risk assessment can be demonstrated by the organisation then there is no reason why such assessment should not be both credible and thus acceptable to clients and regulators.

Clearly, some evidence of external involvement in the setting up and periodic auditing of self-assessment schemes will enhance this credibility provided that the external consultant or organisation can demonstrate sufficient competence in this area.

Proactive involvement in Professional Institutions, Industrial Research Organisations or the production and development of IEC 61508 and associated Standards by both self-assessors and external consultants would assist in this respect. The authors, for example, have made major contributions to the Standard and to a number of the second tier documents described in Chapter 9. Thus, the credibility of third party assessment bodies or consultants does need to be addressed vigorously.

Figure 10.2 shows how a 'DEMONSTRATION OF CONFORMANCE' might be built up from the elements described in this chapter. This 'DEMONSTRATION' would provide backup to any safety report where a level of safety integrity is being

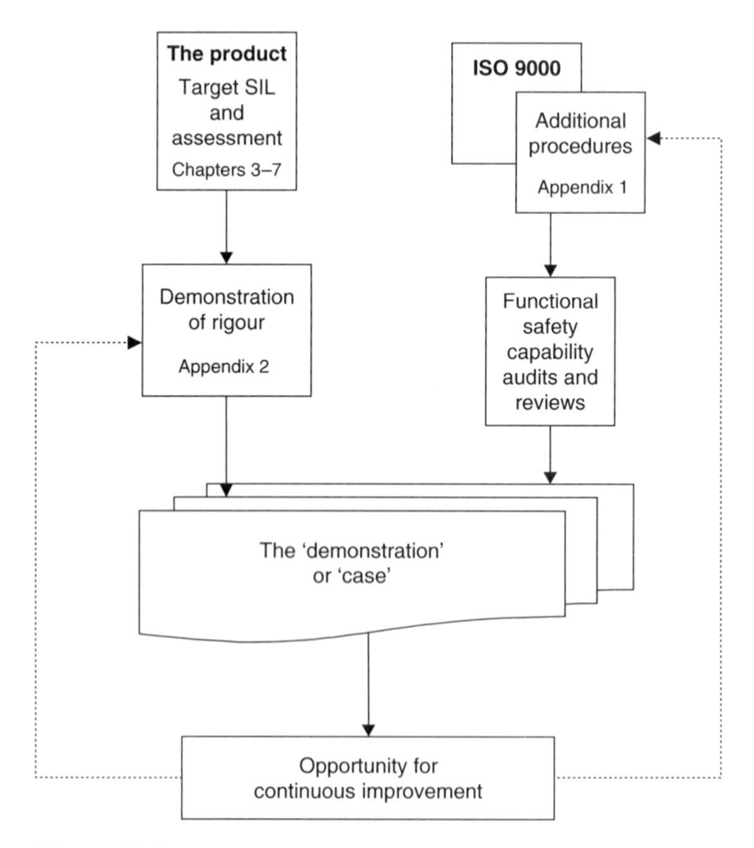

Figure 10.2
Elements of self-assessment

claimed. It also provides a mechanism for continuous improvement as suggested by the assessment techniques themselves.

10.4 Other types of 'certification'

10.4.1 TUV (Germany)

TUV (Technischer Uberwachungs Verein) test houses (in Germany) offer certification of products, against DIN or International Standards. For PLCs in safety-related systems, the certification used to be against DIN V 19250 (see Chapter 9 of this book) and the comparison of SIL definitions with IEC 61508 is shown in Table 10.1.

However, certification is now to IEC 61508.

TUV is not an accredited body for IEC 61508 assessment, in the sense that SIRA is UKAS accredited for Type 5. It therefore provides a 'certification' based on its own credibility and reputation in much the same way as either of the authors of this book would provide a third party assessment.

TUV certification is thus much the same, in principle, as the process described in 10.3 above.

Table 10.1 Comparison of Safety-integrated levels

Probability of failure to function on demand	*Safety-integrity level 61508 and S84.01*	*Class TUV*	*Class DIN V 19250*	*DO-178B S/W class*
$\geqslant 10^{-5}$ to 10^{-4}	4	AK7 & 8*	7	A
$\geqslant 10^{-4}$ to 10^{-3}	3	AK5 & 6	5 & 6	B
$\geqslant 10^{-3}$ to 10^{-2}	2	AK4	4	C
$\geqslant 10^{-2}$ to 10^{-1}	1	AK2 & 3	2 & 3	D
Less than 10^{-1}	NSR	AK1	1	E

*Applies to hardware only

10.4.2 Factory Mutual (USA)

Factory Mutual is an engineering, research and development organisation which offers safety certification against various Standards, including IEC 61508.

They offer certificates of compliance but rightly insist that certification is as ‘fit for use’ in a specific SIL, in a specific application.

The assessments cover:

- Probability of failure on demand
- Test intervals
- Conditions of use of the item in question
- Manufacturer’s safety management system
- Compliance with IEC 61508 requirements
- Common cause assessment
- Safe failure fraction
- FMEA
- Diagnostic coverage assessment
- Environmental testing

Again, this is much the same as described in Section 10.3.

10.5 Preparing for assessment

Whether the assessment is by an accredited body (e.g. SIRA) or a third party consultant, it is important to prepare in advance. The assessor does not know what you know and, therefore, the only visibility of your conformance is provided by documented evidence of:

- Functional safety procedures
- Specifications
- Audits against procedures
- Reviews of the adequacy of procedures
- Design reviews of projects
- Test plans, reports and remedial action
- Safety-integrity assessments

A visible trail of reviews, whereby the procedures and work practices have been developed in practice, is a good indicator that your organisation is committed to functional safety.

Being ill-prepared for an assessment is very cost ineffective. Manhours and fees are wasted on being told what a simple internal audit could have revealed.

The majority of assessments are based on the method of:

- A pre-assessment to ascertain if the required procedures and practice are in place.
- A final assessment where the procedures are reviewed in detail and evidence is sought as to their implementation.

With sensible planning these stages can be prepared for in advance and the necessary reviews conducted internally. It is important that evidence is available to assessors for all the elements of the life-cycle.

Assessments will usually result in:

- Major non-compliances
- Minor non-compliances
- Observations

A major non-compliance would arise if a life-cycle activity is clearly not evidenced. For example, the absence of any requirement for assessment of safe failure fraction would constitute a major non-compliance with the Standard. More than one major non-compliance would be likely to result in the assessment being suspended until the client declared himself ready for reassessment. This would be unnecessarily expensive when the situation could be prevented by adequate preparation.

A minor non-compliance might arise if an essential life-cycle activity, although catered for in the organisation's procedures, has been omitted. For example, a single project where there were inadequate test records would attract a minor non-compliance.

Observations might include comments of how procedures might be enhanced. An example might be the desirability of quoting a proof-test interval in a maintenance manual.

10.6 Summary

It is important to ensure that any assessment concentrates primarily on the technical aspects of a safety-related system in as much as it should address all the aspects (quantitative and qualitative) described in Chapters 2 to 8 of this book.

Procedures and document hierarchies are important, of course, for without them the technical assessment would have no framework upon which to exist and no visibility to demonstrate its findings. However, there is a danger that an 'ISO 9000 mentality' approach can concentrate solely on the existence of procedures and of specific document titles. Procedures, and the mere existence of documents, do not of themselves imply achieved functional safety unless they result in technical activity. Similarly documents do not create safety, they are a vehicle to implement technical requirements. Their titles are relatively unimportant and it is necessary to see behind them to assess whether the actual requirements described in this book have been addressed and implemented. The same applies to safety management systems generally.

If this is borne in mind then assessment, be it self-generated or third party, can be highly effective.

PART E

CASE STUDIES IN THE FORM OF EXERCISES AND EXAMPLES

In this section there are two case studies, the first of which is in the form of an exercise.

Chapter 11 is an exercise involving the selection of a target SIL for a pressure let down system. The design is compared with the target and improvements are evaluated and subjected to ALARP criteria. The answers are provided in Appendix 5.

Chapter 12 is a typical assessment report on a burner control system. The reader can compare and critique this, having read the earlier chapters of this book.

Chapter 13 presents a number of rather different SIL targeting examples.

Chapter 14 is a purely hypothetical proposal for a rail train braking system.

These case studies address the four quantitative aspects of IEC 61508:

- SIL targeting
- Random hardware failures
- Safe failure fraction
- ALARP

CHAPTER 11

PRESSURE CONTROL SYSTEM (EXERCISE)

This exercise is based on a real scenario. Spaces have been left for the reader to attempt the calculations. The answers are provided in Appendix 5.

11.1 The unprotected system

Consider a plant supplying a gas to offsite via a twin stream pressure control station. Each stream is regulated by two valves (top of Figure 11.1). Each valve is under the control of its downstream pressure. Each valve is closed by the upstream gas pressure via its pilot valve, J, but only when its pilot valve, K1, is closed. Opening pilot valve K1 relieves the pressure on diaphragm of valve, V, allowing it to open. Assume that a HAZOP (HAZard and OPerability) study of this system establishes that downstream overpressure, whereby the valves fail to control the downstream pressure, is an event which could lead to one or more fatalities.

Since the risk is offsite a target maximum tolerable risk of 10^{-5} per annum has been proposed.

Assume that a quantified risk assessment has predicted a probability of 20% that failure, involving overpressure, will lead to subsequent pipe rupture and ignition. Furthermore it is predicted that, due to the high population density, fatality is 50% likely.

Assume also that the plant offers approximately 10 risks in total to the same population (e.g. tanker deliveries, other pipelines, site explosion).

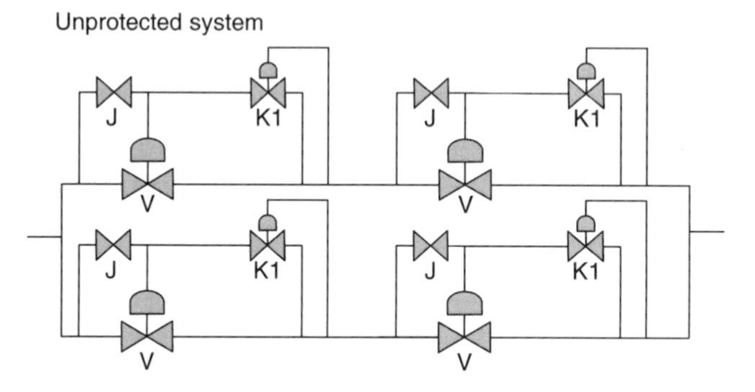

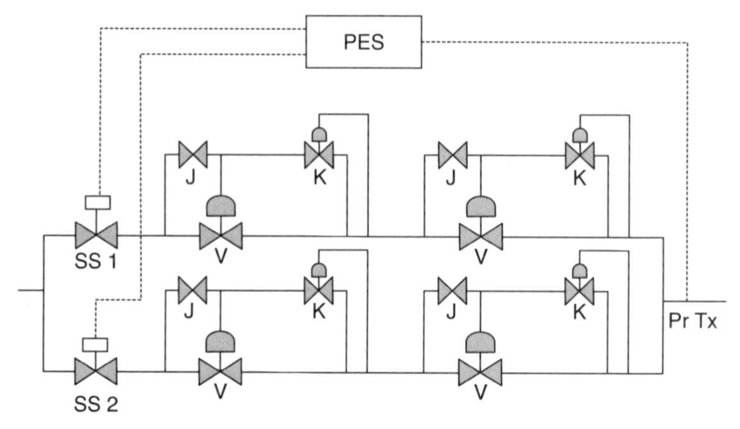

Figure 11.1
The system, with and without backup protection

It follows that the target failure rate for overpressure of the twin stream sub-system is 10^{-5}/[10 risks $\times$ 0.2 $\times$ 0.5] = $\mathbf{10^{-5}}$**pa**.

Assume, however, that field experience of a significant number of these twin stream systems shows that the frequency of overpressure is dominated by the pilots and is $\mathbf{2.5 \times 10^{-3}}$**pa**.

11.2 Protection system

Since 2.5×10^{-3} is greater than 10^{-5} a design modification is proposed whereby a programmable electronic system (PES) closes a valve in each stream, based on an independent measure of the downstream pressure. The valves consist of actuated ball valves (sprung to close). This is illustrated at the bottom of Figure 11.1.

The target Unavailability for this 'add-on' safety system is therefore ...... which indicates an SIL of ......

11.3 Assumptions

The following assumptions are made in order to construct and quantify the reliability model:

(a) Failure rates (symbol λ), for the purpose of this prediction, are assumed to be constant with time. Both early and wearout related failures are assumed to be removed by burn-in and preventive replacement respectively.
(b) The MTTR (Mean Time To Repair) of a revealed failure is 4 hours.
(c) The auto-test coverage of the PLC is 90% and occurs at just under 5 minute intervals. The MDT (Mean Down Time) for failures revealed by this PES auto-test are taken to be the same as the MTTR (Mean Time To Repair) because the MTTR >> the auto-test period. The MDT is thus assumed to be 4 hours. Neither the pressure transmitter nor the valve is assumed to have any self-diagnostics.
(d) The manual proof test is assumed to be 100% effective and to occur annually (c. 8000 hours).
(e) One maintenance crew is assumed to be available for each of the three equipment types (PES, Instrumentation, Pneumatics).
(f) The detailed design assumptions needed for an assessment of the common cause failure BETA factor (see modified proposal) are summarised in Section 11.8.

11.4 Reliability block diagram

Figure 11.2 is the reliability block diagram for the add-on safety system. Note that the PES will occur twice in the diagram. This is because the model needs to address those failures revealed by auto-test separately from those revealed by the longer manual proof test due to their different MDTs (explained more fully in Section 6.3).

TO BE FILLED IN BY THE READER
(See Appendix 5 for answer)

Figure 11.2
Reliability block diagram

11.5 Failure rate data

The following failure rate data will have been chosen for the protection system components, shown in Figure 11.1. These are the component level failure modes which lead to the hazard under consideration (i.e. downstream overpressure). FARADIP.THREE has been used to obtain the failure rates.

Item	*Failure mode*	*Failure rates 10^{-6} per hour* (total)	(mode)
PES	PES low or zero*	5	0.25
Pressure transmitter	Fail low	2	0.5 (25% has been assumed)
Actuated ball valve (sprung to close)	Fail to close	8	0.8**

* This represents any failure of the PES i/p, CPU or o/p causing the low condition
** 10% has been used based on the fact that the most likely failure mode is fail closed

11.6 Quantifying the model

The following Unavailability calculations address each of the groups (left to right) in Figure 11.2 (see Appendix 5):

(a) Ball valve 1 – unrevealed failures
Unavailability = ...
= ......

(b) Ball valve 2 – unrevealed failures
Unavailability = ...
= ......

(c) PES output 1 failures revealed by auto-test
Unavailability = ...
= ......

(d) PES output 1 failures not revealed by auto-test
Unavailability = ...
= ......

(e) PES output 2 failures revealed by auto-test
Unavailability = ...
= ......

(f) PES output 2 failures not revealed by auto-test
Unavailability = ...
= ......

(g) Pressure transmitter – unrevealed failures
Unavailability = ...
= ......

The predicted Unavailability is obtained from the sum of the unavailabilities in (a) to (g) = ...

11.7 Proposed design and maintenance modifications

The proposed system is not acceptable (as can be seen in Appendix 5) and modifications are required.

Before making modification proposals it is helpful to examine the relative contributions to system failure of the various elements in Figure 11.2.

TO BE FILLED IN BY THE READER
(See Appendix 5 for answer)

Figure 11.3
Revised reliability block diagram (or fault tree)

??% from items (a) and (b) ball valve.
??% from items (c) to (f) the PES.
??% from item (g) the pressure transmitter.

It was decided to duplicate the pressure transmitter and vote the pair (1 out of 2). It was also decided to reduce the proof test interval to six months (c. 4000 hours).

11.8 Modelling common cause failure (pressure transmitters)

The BETAPLUS method provides a method for assessing the percentage of common cause failures. The scoring for the method was carried out assuming:

- Written procedures for system operation and maintenance are evident but not extensive.
- There is some training of all staff in CCF awareness.
- Extensive environmental testing was conducted.
- Identical (i.e. non-diverse) redundancy.
- Basic top level FMEA (failure mode analysis) had been carried out.

- There is some limited field failure data collection.
- Simple, well-proven, pressure transmitters 1/2 metre apart with cables routed together.
- Good electrical protection.
- Annual proof test.

The BETAPLUS software package performs the calculations and was used to calculate a BETA value of 9%.

11.9 Quantifying the revised model

The following takes account of the pressure transmitter redundancy, common cause failure and the revised proof-test interval. Changed figures are shown in bold in Appendix 5.

(a) Ball valve SS1 fails open.
Unavailability = ...
= ......

(b) Ball valve SS2 fails open.
Unavailability = ...
= ......

(c) PES output 1 fails to close valve (undiagnosed failure).
Unavailability = ...
= ......

(d) PES output 2 fails to close valve (undiagnosed failure).
Unavailability = ...
= ......

(e) PES output 1 fails to close valve (diagnosed failure).
Unavailability = ...
= ......

(f) PES output 2 fails to close valve (diagnosed failure).
Unavailability = ...
= ......

(g) Voted pair of pressure transmitters.
Unavailability = ...
= ......

(h) Common cause failure of pressure transmitters.
Unavailability = ...
= ......

The predicted Unavailability is obtained from the sum of the unavailabilities in (a) to (h) = ...

11.10 ALARP

Assume that further improvements in CCF can be achieved for a total cost of £1000. Assume, also, that this results in an improvement in unavailability to $\mathbf{4 \times 10^{-4}}$. It is necessary to consider, applying the ALARP principle, whether this improvement should be implemented.

The cost per life saved over a 40-year life of the equipment (without cost discounting) is calculated, assuming two fatalities, as follows:

?????? (see Appendix 5)

11.11 Architectural constraints

Consider the architectural constraints imposed by IEC 61508 Part 2, outlined in Section 3.3.2.

Do the pressure transmitters and valves in the proposed system meet the minimum architectural constraints assuming they are 'TYPE A components'?

Does the PES, in the proposed system, meet the minimum architectural constraints assuming it is a 'TYPE B component'?

CHAPTER 12

BURNER CONTROL ASSESSMENT (EXAMPLE)

This chapter consists of a possible report of an integrity study on a proposed replacement burner control system. Unlike Chapter 11, the requirement involves the high demand table and the target is expressed as a failure rate.

This is not intended as a MODEL report but an example of a typical approach. The reader may care to study it in the light of this book and attempt to list omissions and to suggest improvements.

SAFETY INTEGRITY STUDY OF A PROPOSED REPLACEMENT BOILER CONTROLLER

CONTENTS

Issue 1.0
1 February 2001

EXECUTIVE SUMMARY & RECOMMENDATIONS

Objectives

To establish a Safety-Integrity Level target, *vis-à-vis* IEC 61508, for a Boiler Control System which is regarded as safety-related. To address the following failure mode: Pilots are extinguished but nevertheless burner gas continues to be released with subsequent explosion of the unignited gas. To assess the design against the above target and to make recommendations.

Targets

A Maximum Tolerable Risk target of 10^{-4} per annum which leads to a MAXIMUM TOLERABLE TARGET FAILURE RATE of 3×10^{-3} per annum (see Section 12.2).

This implies a SIL 2 target.

Results

The frequency of the top event is 2×10^{-4}pa and the target is met. This result remains within the ALARP region but it was shown that further risk reduction is unlikely to be justified.

Recommendations

Review all the assumptions in Sections 12.2, 12.3 and 12.4.3. Review the failure rates and down times in Section 12.5 and the fault tree logic, in Figures 12.1–12.3, for a future version of this study.

Continue to address ALARP.

Place a SIL 2 requirement on the system vendor, in respect of the requirements of Parts 2 and 3 of IEC 61508.

Because very coarse assumptions have had to be made, concerning the PLC and SAM (safety monitor) design, carry out a more detailed analysis with the chosen vendor.
Address the following design considerations with the vendor:

- Effect of loss of power supply, particularly where it is to only some of the equipment.

- Examine the detail of the PLC/SAM interconnections to the I/O and ensure that the fault tree logic is not compromised.
- Establish if the effect of failure of the valve limit switches needs to be included in the fault tree logic.

12.1 Objectives

(a) To establish a Safety-Integrity Level target, *vis-à-vis* IEC 61508, for a Boiler Control System which is regarded as safety-related.

(b) To address the following failure mode:
Pilots are extinguished but nevertheless burner gas continues to be released with subsequent explosion of the unignited gas.

(c) To assess the design against the above target.

(d) To make recommendations.

12.2 Integrity requirements

I Gas E SR/15 Amendment suggests target maximum tolerable risk criteria (Table 3.1 of this book). These are, for individual risk:

1–5 FATALITIES (EMPLOYEE)	10^{-4} pa
BROADLY ACCEPTABLE	10^{-6} pa

Assume that there is a 0.9 probability of ignition of the unburnt gases.
Assume that there is a 0.1 probability of the explosion leading to fatality.
Assume that there is a 0.5 probability that the oil burners are not active.
Assume that there is a 0.75 probability of there being a person at risk.

Hence the MAXIMUM TOLERABLE TARGET FAILURE RATE = 10^{-4} pa divided by $(0.9 \times 0.1 \times 0.5 \times 0.75)$

$$= 3 \times 10^{-3} \text{ per annum}$$

This invokes a SIL 2 target.

12.3 Assumptions

12.3.1 Specific

(a) Proof test is carried out annually. Thus the mean down time of unrevealed failures, being half the proof-test interval, is approximately 4000 hours.

(b) The system is in operation 365 days per annum.

(c) The burner control system comprises a combination of four 'XYZ Ltd' PLCs and a number of safety monitors (known as SAMs).

12.3.2 General

(a) Reliability assessment is a statistical process for applying historical failure data to proposed designs and configurations. It therefore provides a credible target/estimate of the likely reliability of equipment assuming manufacturing, design and operating conditions identical to those under which the data were collected. It is a valuable design review technique for comparing alternative designs, establishing order of magnitude performance targets and evaluating the potential effects of design changes.

(b) Failure rates (symbol λ), for the purpose of this prediction, are assumed to be constant with time. Both early and wearout related failures would decrease the reliability but are assumed to be removed by burn-in and preventive replacement respectively.

(c) Each single component failure which causes system failure is described as a SERIES ELEMENT. This is represented, in fault tree notation, as an OR gate whereby any failure causes the top event. The system failure rate contribution from this source is obtained from the sum of the individual failure rates.

(d) Where coincident failures are needed to fail for the relevant system failure mode to occur then this is represented, in fault tree notation, as an AND gate where more than one failure is needed to cause the top event.

(e) The failure rates used, and thus the predicted MTBFs (mean time between failure) and availabilities, are those credibly associated with a well proven design after a suitable period of reliability growth. They might therefore be considered optimistic as far as field trial or early build states are concerned.

(f) Calendar based failure rates have been used in this study.

(g) Software failures are systematic and, as such, are not random. They are not quantified in this study.

12.4 Results

12.4.1 Random hardware failures

The fault tree logic was constructed from a discussion of the failure scenarios at the meeting on 8 January 2001 involving Messrs 'Q' and 'Z'. The fault tree was analysed using the TECHNIS fault tree package TTREE.

The frequency of the top event (Figure 12.1) is 2×10^{-4}pa (see Annex 1) which is well within the target.

Annex 1 shows the combinations of failures (cut sets) which lead to the failure mode in question. It is useful to note that at least three coincident events are required to lead to the top

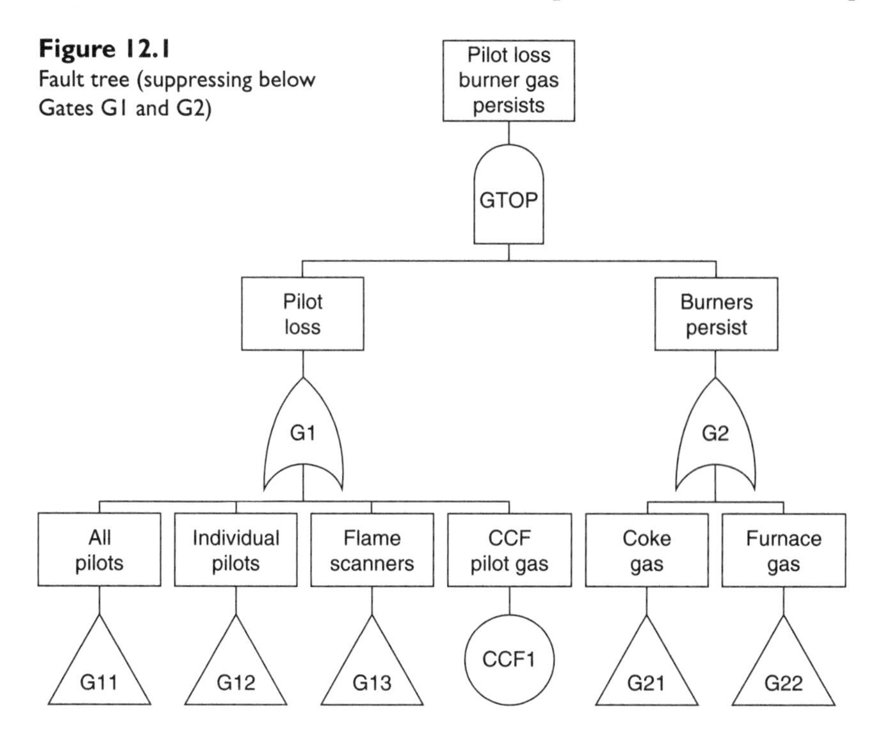

Figure 12.1
Fault tree (suppressing below Gates G1 and G2)

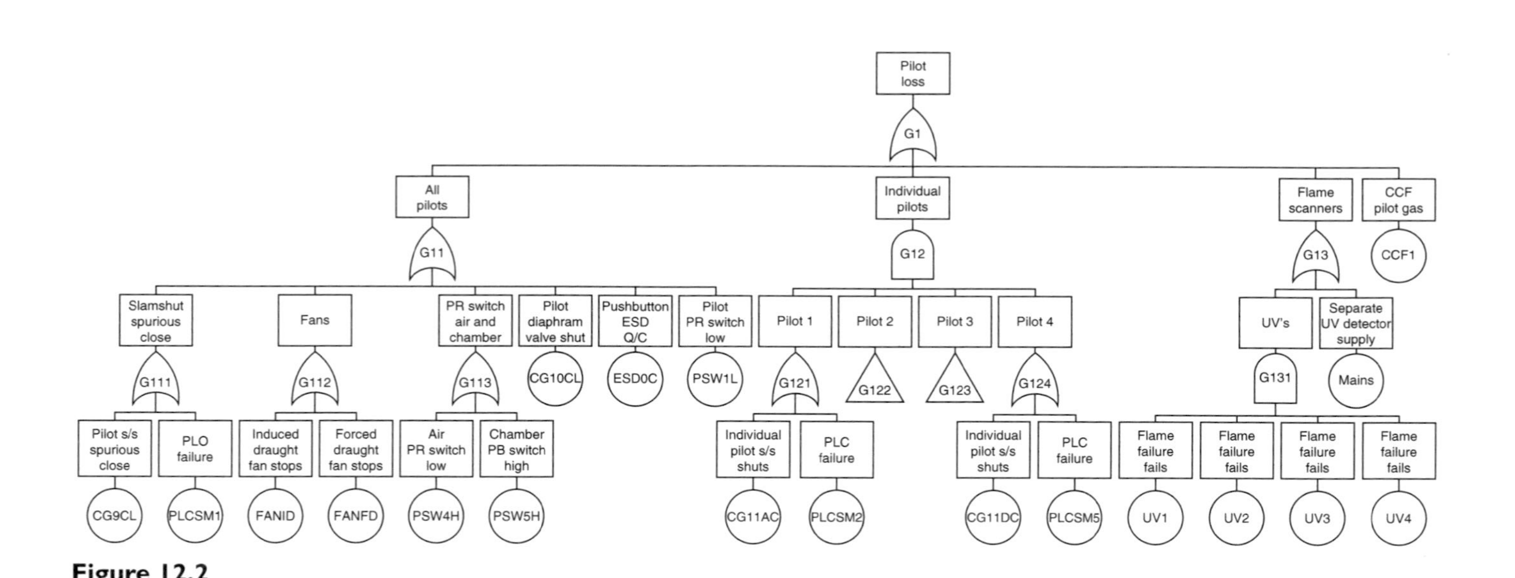

Figure 12.2
Fault tree (Gate G1)

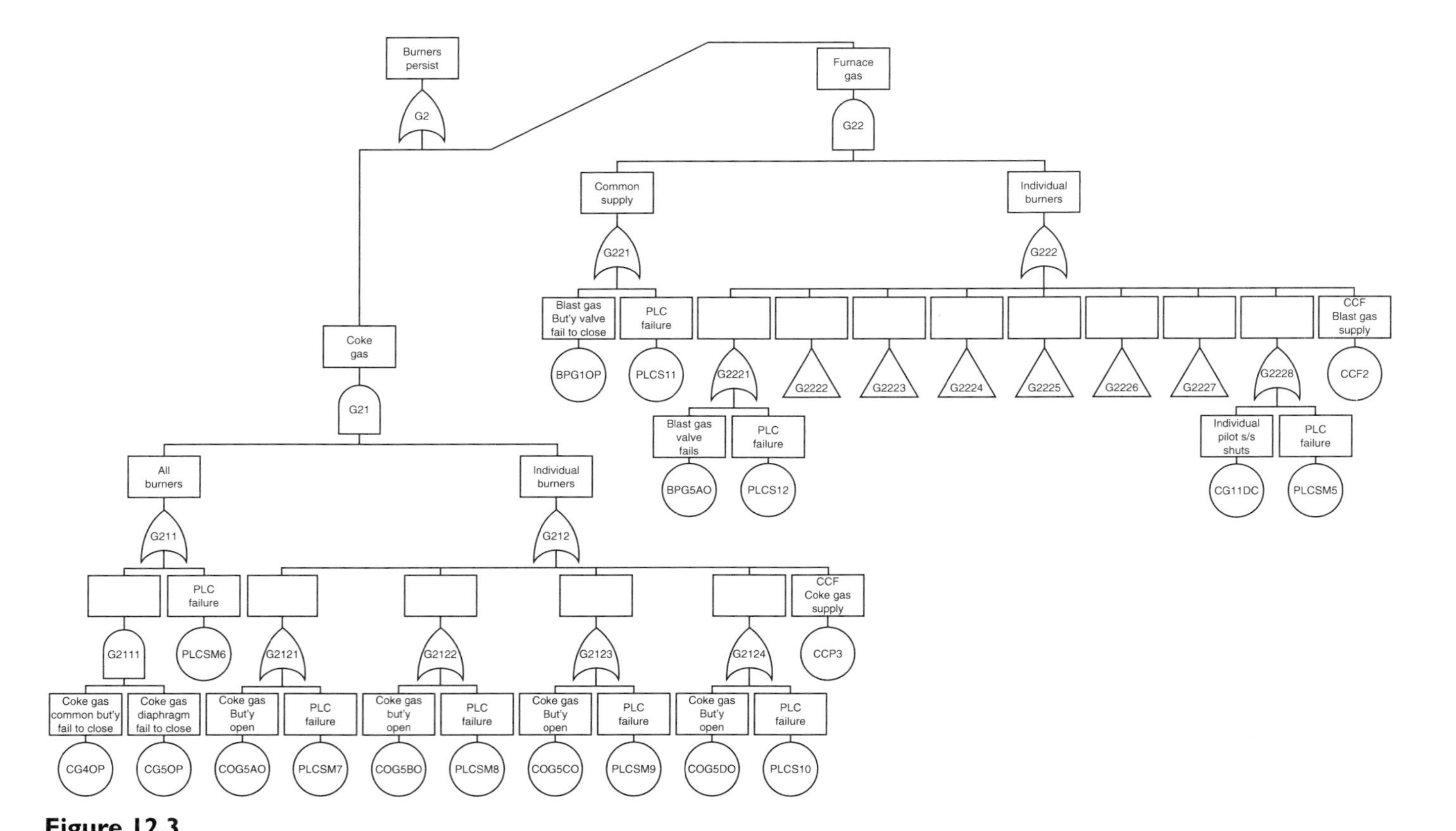

Figure 12.3
Fault tree (Gate G2)

event. An 'Importance' measure is provided for each cut set and it can be seen that no cut set contributes more than 1.4% of the total. There is therefore no suggestion of a critical component.

12.4.2 Qualitative requirements

The qualitative measures required to limit software failures are listed, for each SIL, in the I Gas E SR/15 and IEC 61508 documents. Although the I Gas E guidance harmonises closely with IEC 61508, compliance with SR/15 does not automatically imply compliance with IEC 61508.

It has to be stressed that this type of qualitative assessment merely establishes a measure of 'adherence to a process' and does not signify that the quantitative SIL is automatically achieved by those activities. It addresses, however, a set of measures deemed to be appropriate (at the SIL) by the above documents.

It should also be kept in mind that an assessment is in respect of the specific failure mode. The assessment of these qualitative measures should therefore, ideally, be in respect of their application to those failure modes rather than in a general sense.

THE PURPOSE OF THE FOLLOWING IS TO PROVIDE AN *AIDE-MEMOIRE* WHEREBY FEATURES OF THE DESIGN CYCLE CAN BE ASSESSED IN GREATER DETAIL FOR INCLUSION IN A LATER ASSESSMENT. THIS LIST IS BASED ON SAFETY-INTEGRITY LEVEL (SIL 2).

1 Requirements

(a) Requirements Definition: This needs to be identified. It needs to be under configuration control with adequate document identification. It should also refer to the safety-integrity requirements of the failure mode addressed in this report. *Subject to this, the requirement will be met.* A tender document, in response to the Requirements Specification, might well have been produced by the supplier and might well be identified.

(b) The Functional Specification needs to address the safety-integrity requirement and to be specific about the failure modes. It will be desirable to state to the client that it is understood that the integrity issue is 'loss of pilot followed by ...' etc. *Subject to this, the requirement will be met.*

(c) The design may not utilise a CAD specification tool or formal method in delineating the requirement. However, the safety-related system might comprise simple control loops and therefore not involve parameter calculation, branching decision algorithms or complex data manipulation. Thus, a formal specification language may not be applicable. The documentation might be controlled by ISO 9001 configuration control and appropriate software management. The need for an additional CAD specification tool may not be considered necessary. *Subject to this, the requirement will be met.*

2 Design and language

(a) There should be evidence of a 'structured' design method. Examples include:

Logic diagrams
Data dictionary
Data flow diagrams
Truth tables

Subject to this, the requirement will be met.

(b) There should be a company specific, or better still, project specific coding/design standard which addresses, for example (list where possible):

Use of a suitable language
Compiler requirements
Hygienic use of the language set
Use of templates (i.e. field proven) modules
No dynamic objects
No dynamic variables or online checking thereof
Limited interrupts, pointers and recursion
No unconditional jumps
Fully defined module interfaces

Subject to this, the requirement will be met.

(c) Ascertain if the compiler/translator certified or internally validated by long use. *Subject to this, the requirement will be met.*

(d) Demonstrate a modular approach to the structure of the code and rules for modules (i.e. single entry/exit). *Subject to this, the requirement will be met.*

3 Fault tolerance

(a) Assuming Type B components, and a non-redundant configuration, at least 90% safe failure fraction is required for SIL 2. It will be necessary to establish that 90% of PLC failures are either detected by the watch-dog or result in failures not invoking the failure mode addressed in this study. *Subject to a review the requirement will be met.*

(b) Desirable features (not necessarily essential) would be, error detection/correction codes and failure assertion programming. *Subject to this, the requirement will be met.*

(c) Demonstrate graceful degradation in the design philosophy. *Subject to this, the requirement will be met.*

4 Documentation and change control

(a) A description is needed here to cover: Rigour of configuration control (i.e. document master index, change control register, change notes, change procedure, requirements matrix (customer spec/FDS/ FAT mapping). *Subject to this, the requirement will be met.*

(b) The change/modification process should be fairly rigorous, key words are:

Impact analysis of each change
Re-verification of changed and affected modules (the full test not just the perceived change)
Re-verification of the whole system for each change
Data recording during these re-tests

Subject to this, the requirement will be met.

5 Design review

(a) Formal design review procedure? Evidence that design reviews are:

Specifically planned in a Quality Plan document

Which items in the design cycle are to be reviewed (i.e. FDS, acceptance test results etc.)
Described in terms of who is participating, what is being reviewed, what documents etc.
Followed by remedial action
Specifically addressing the above failure mode
Code review see (b)

Subject to this, the requirement will be met.

(b) Code: Specific code review at pseudo code or ladder or language level which addresses the above failure mode. *Subject to this, the requirement will be met.*

(c) There needs to be justification that the language is not suitable for static analysis and that the code walkthrough is sufficiently rigorous for a simple PLC language set in that it is a form of 'low level static analysis'. *Subject to this, the requirement will be met.*

6 Test (applies to both hardware and software)

(a) There should be a comprehensive set of functional and interface test procedures which address the above failure mode. The test procedures need to evidence some sort of formal test case development for the software (i.e. formally addressing the execution possibilities, boundary values and extremes). *Subject to this, the requirement will be met.*

(b) There should be mis-use testing in the context of failing due to some scenario of I/O or operator interface. *Subject to this, the requirement will be met.*

(c) There should be evidence of formal recording and review of all test results including remedial action (probably via the configuration and change procedures). *Subject to this, the requirement will be met.*

(d) There should be specific final validation test plan for proving the safety-related feature. This could be during commissioning. *Subject to this, the requirement will be met.*

7 Integrity assessment

Reliability modelling has been used in the integrity assessment.

8 Quality, safety and management

(a) In respect of the safety-integrity issues (i.e. for the above failure mode) some evidence of specific competency mapping to show that individuals have been chosen for tasks with the requirements in view (e.g. safety testing, integrity assessment). The competency requirements of IEC 61508 infer that appropriate job descriptions and training records for operating and maintenance staff are in place. *Subject to this, the requirement will be met.*

(b) Show that an ISO 9001 quality system is in operation, if not actually certified. *Subject to this, the requirement will be met.*

(c) Show evidence of safety management in the sense of ascertaining safety engineering requirements in a project as is the case in this project. This study needs to address the safety management system (known as functional safety capability) of the equipment designer and operator. Conformance with IEC 61508 involves this aspect of the safety-related equipment. *Subject to this, the requirement will be met.*

(d) Failure recording, particularly where long term evidence of a component (e.g. the compiler or the PLC hardware) can be demonstrated is beneficial. *Subject to this, the requirement will be met.*

9 Installation and commissioning

There needs to be a full commissioning test. Also, modifications will need to be subject to control and records will need to be kept. *Subject to this, the requirement will be met.*

12.4.3 ALARP

The ALARP (as low as reasonably practicable) principle involves deciding if the cost and time of any proposed risk reduction is, or is not, grossly disproportionate to the safety benefit gained.

The demonstration of ALARP is supported by calculating the Cost per Life Saved of the proposal. The process is described in Chapter 3. Successive improvements are considered in this fashion until the cost becomes disproportionate. The target of

3×10^{-3}pa corresponded to a maximum tolerable risk target of 10^{-4}pa. The resulting 2×10^{-4}pa corresponds to a risk of 6.6×10^{-6}pa. This individual risk is not as small as the BROADLY ACCEPTABLE level and ALARP should be considered.

Assuming, for the sake of argument, that the scenario is sufficiently serious as to involve two fatalities then any proposed further risk reduction would need to be assessed against the ALARP principle. Assuming a £2 000 000 per life saved criterion then the following would apply to a proposed risk reduction, from 6.6×10^{-6}pa. Assuming a 30-year plant life:

$$\pounds 2\,000\,000 = \frac{\text{(Proposed expenditure)}}{([6.6 \times 10^{-6} - 10^{-6}] \times 30 \times 2)}$$

Thus: proposed expenditure = £672

It seems unlikely that the degree of further risk reduction referred to could be achieved within £672 and thus it might be argued that ALARP is satisfied.

12.5 Failure rate data

In this study the FARADIP.THREE Version 4.1 data ranges have been used for some of the items. The data are expressed as ranges. In general the lower figure in the range, used in a prediction, is likely to yield an assessment of the credible design objective reliability. That is the reliability which might reasonably be targeted after some field experience and a realistic reliability growth programme. The initial (field trial or prototype) reliability might well be an order of magnitude less than this figure. The centre column figure (in the FARADIP software package) indicates a failure rate which is more frequently indicated by the various sources. It has been used where available. The higher figure will probably include a high proportion of maintenance revealed defects and failures. F3 refers to FARADIP.THREE, Judge refers to judgement.

Code (Description)	Mode	Failure rate PMH (or fixed probability)	Mode rate 10^{-6} per hour	MDT (hrs)	Reference
CCF1 (Common Cause Failures)	any	0.1	0.1	24	JUDGE
CCF2/3 (Common Cause Failures)	any	0.1	0.1	4000	JUDGE
ESDOC (ESD button)	o/c	0.1	0.1	24	F3
UV (UV detector)	fail	5	2	24	F3
MAINS (UV separate supply)	fail	5	5	24	JUDGE
PLC… (Revealed failures)	–	5	1	24	JUDGE
PLC… (Unrevealed failures)	–	5	1	4000	JUDGE
FAN (Any fan)	fail	10	10	24	F3
PSWL (Pressure switch)	low	2	1	24	F3
PSWH (Pressure switch)	high	2	1	24	F3
CG10CL (Pilot diaphragm vlv)	closed	2	1	24	F3
CG9CL (Slamshut)	sp close	–	1	24	F3
CG11… (Slamshuts)	sp close	–	4	24	F3
COG5… (Butterfly vlv)	fail to close	–	2	4000	F3
CG4OP… (Butterfly vlv)	fail to close	–	2	4000	F3
CG5OP (Diaphragm vlv)	fail to close	–	2	4000	F3
BFG… (Blast gas vlvs)	–	–	2	4000	F3

12.6 References

A reference section would normally be included.

Annex I Fault tree details

File name: Burner.TRO

Results of fault tree quantification for top event: GTOP

Top event frequency = 0.222E−07 per hour
= 0.194E−03 per year

Top event MTBF = 0.451E+08 hours
= 0.515E+04 years

Top event probability = 0.526E−06

Basic event reliability data

Basic event	*Type*	*Failure rate*	*Mean fault duration*
CCF1	I/E	0.100E−06	24.0
CG10CL	I/E	0.100E−05	24.0
ESDOC	I/E	0.100E−06	24.0
PSW1L	I/E	0.100E−05	24.0
CG9CL	I/E	0.100E−05	24.0
PLCSM1	I/E	0.100E−05	24.0
FANID	I/E	0.100E−04	24.0
FANFD	I/E	0.100E−04	24.0
PSW4H	I/E	0.100E−05	24.0
PSW5H	I/E	0.100E−05	24.0
CG11AC	I/E	0.400E−05	24.0
PLCSM2	I/E	0.100E−05	24.0
CG11BC	I/E	0.400E−05	24.0
PLCSM3	I/E	0.100E−05	24.0
CG11CC	I/E	0.400E−05	24.0
PLCSM4	I/E	0.100E−05	24.0
CG11DC	I/E	0.400E−05	24.0
PLCSM5	I/E	0.100E−05	24.0
MAINS	I/E	0.500E−05	24.0
UV1	I/E	0.200E−05	24.0
UV2	I/E	0.200E−05	24.0
UV3	I/E	0.200E−05	24.0
UV4	I/E	0.200E−05	24.0

Basic event	*Type*	*Failure rate*	*Mean fault duration*
PLCSM6	I/E	0.100E−05	0.400E+04
CCF3	I/E	0.100E−06	0.400E+04
COG5AO	I/E	0.200E−05	0.400E+04
PLCSM7	I/E	0.100E−05	0.400E+04
COG5BO	I/E	0.200E−05	0.400E+04
PLCSM8	I/E	0.100E−05	0.400E+04
COG5CO	I/E	0.200E−05	0.400E+04
PLCSM9	I/E	0.100E−05	0.400E+04
COG5DO	I/E	0.200E−05	0.400E+04
PLCS10	I/E	0.100E−05	0.400E+04
CG4OP	I/E	0.200E−05	0.400E+04
CG5OP	I/E	0.200E−05	0.400E+04
BFG1OP	I/E	0.100E−05	0.400E+04
PLCS11	I/E	0.100E−05	0.400E+04
CCF2	I/E	0.100E−06	0.400E+04
BFG5AO	I/E	0.100E−05	0.400E+04
PLCS12	I/E	0.100E−05	0.400E+04
BFG5BO	I/E	0.100E−05	0.400E+04
PLCS13	I/E	0.100E−05	0.400E+04
BFG5CO	I/E	0.100E−05	0.400E+04
PLCS14	I/E	0.100E−05	0.400E+04
BFG5DO	I/E	0.100E−05	0.400E+04
PLCS15	I/E	0.100E−05	0.400E+04
BFG5EO	I/E	0.100E−05	0.400E+04
PLCS16	I/E	0.100E−05	0.400E+04
BFG5FO	I/E	0.100E−05	0.400E+04
PLCS17	I/E	0.100E−05	0.400E+04
BFG5GO	I/E	0.100E−05	0.400E+04
PLCS18	I/E	0.100E−05	0.400E+04
BFG5HO	I/E	0.100E−05	0.400E+04
PLCS19	I/E	0.100E−05	0.400E+04

Barlow-Proschan measure of cut set importance (Note: This is the name given to the practice of ranking cut sets by frequency)

Rank 1 Importance 0.144E−01 MTBF hours 0.313E+10 MTBF years 0.357E+06

Basic event	*Type*	*Failure rate*	*Mean fault duration*
FANID	I/E	0.100E−04	24.0
PLCSM6	I/E	0.100E−05	0.400E+04
COG5AO	I/E	0.200E−05	0.400E+04

Rank 2 Importance 0.144E−01 MTBF hours 0.313E+10 MTBF years 0.357E+06

Basic event	*Type*	*Failure rate*	*Mean fault duration*
FANID	I/E	0.100E−04	24.0
PLCSM6	I/E	0.100E−05	0.400E+04
COG5BO	I/E	0.200E−05	0.400E+04

Rank 3 Importance 0.144E−01 MTBF hours 0.313E+10 MTBF years 0.357E+06

Basic event	*Type*	*Failure rate*	*Mean fault duration*
FANID	I/E	0.100E−04	24.0
PLCSM6	I/E	0.100E−05	0.400E+04
COG5CO	I/E	0.200E−05	0.400E+04

Rank 4 Importance 0.144E−01 MTBF hours 0.313E+10 MTBF years 0.357E+06

Basic event	*Type*	*Failure rate*	*Mean fault duration*
FANID	I/E	0.100E−04	24.0
PLCSM6	I/E	0.100E−05	0.400E+04
COG5DO	I/E	0.200E−05	0.400E+04

Rank 5 Importance 0.144E−01 MTBF hours 0.313E+10 MTBF years 0.357E+06

Basic event	*Type*	*Failure rate*	*Mean fault duration*
FANFD	I/E	0.100E−04	24.0
PLCSM6	I/E	0.100E−05	0.400E+04
COG5AO	I/E	0.200E−05	0.400E+04

Rank 6 Importance 0.144E−01 MTBF hours 0.313E+10 MTBF years 0.357E+06

Basic event	*Type*	*Failure rate*	*Mean fault duration*
FANFD	I/E	0.100E−04	24.0
PLCSM6	I/E	0.100E−05	0.400E+04
COG5BO	I/E	0.200E−05	0.400E+04

CHAPTER 13

SIL TARGETING – SOME PRACTICAL EXAMPLES

13.1 A problem involving EUC/SRS independence

Figure 13.1 shows the same EUC as was used in Chapter 11. In this case, however, the additional protection is provided by means of additional K2 pilot valves, provided for each valve, V. This implies that failure of the valves, V, was (wrongly) not perceived to be significant. Closing the K2 pilot valve (via the PES and an I/P converter) has the same effect as closing the K1 pilot. The valve, 'V', is thus closed by *either* K1 *or* K2. This additional safety-related protection system (consisting of PES, I/P converters and K2 pilots) provides a backup means of closing valve 'V'.

The PES receives a pressure signal from the pressure transmitters P. A 'high' signal will cause the PES to close the K2 pilots and thus valves 'V'.

It might be argued that the integrity target for the add-on SRS (consisting of PESs, transmitters and pilots) is assessed as in Chapter 11. This would lead to the same SIL target as is argued in Chapter 11, namely **2.5×10^{-3} PFD being SIL 2**.

However, there are two reasons why the SRS is far from INDEPENDENT of the EUC:

(a) Failures of the valve V actuators, causing the valves to fail open, will not be mitigated by the K2 pilot.
(b) It is credible that the existing pilots K1 and the add-on pilots K2 might have common cause failures. In that case some failures of K1 pilots would cause failure of their associated K2 pilots.

Therefore, in Chapter 11, a design is offered which does provide EUC/SRS independence. What then of the SIL target for the SRS in Figure 13.1.

It becomes necessary to regard the whole of the system as a single safety-related system. It thus becomes a high demand system with a Maximum Tolerable Failure Rate (see Chapter 11) of 10^{-5}pa. This is at the far limit of SIL 4 and is, of course, quite unacceptable. Thus an alternate design would be called for.

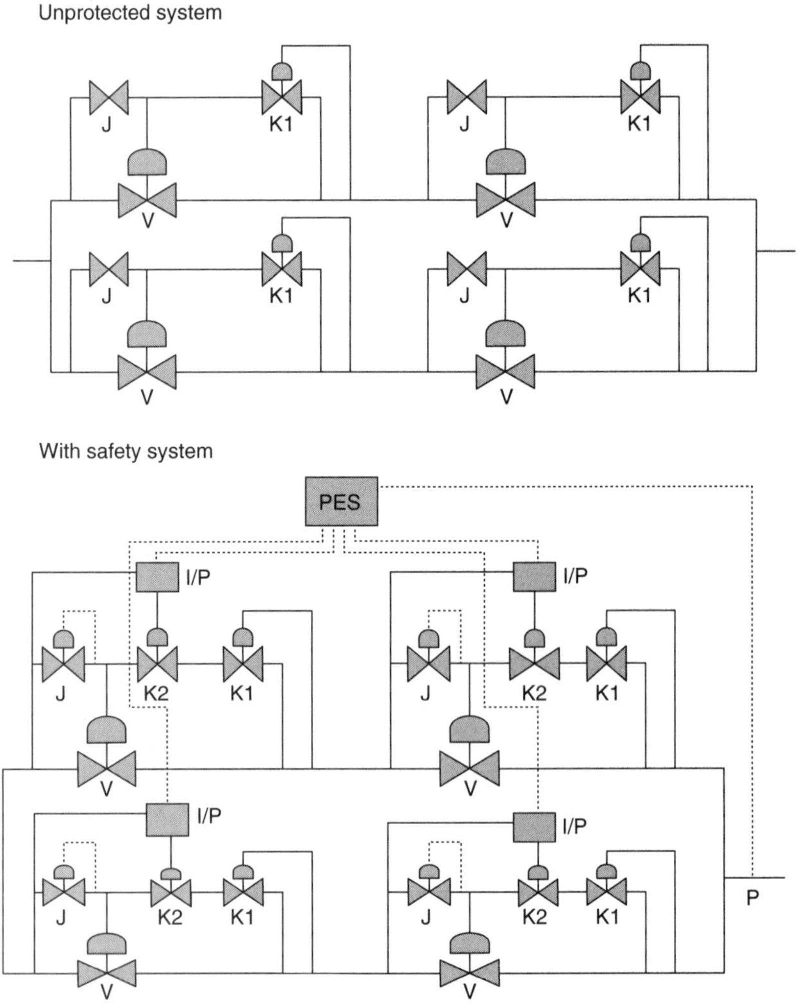

Unprotected system

J K1 J K1 V V J K1 J K1 V V

With safety system

PES I/P I/P J K2 K1 J K2 K1 V V I/P I/P J K2 K1 J K2 K1 P V V

Figure 13.1
The system, with and without backup protection

13.2 A handheld alarm intercom, involving human error in the mitigation

A rescue worker, accompanied by a colleague, is operating in a hazardous environment. The safety-related system, in this example, consists of a handheld intercom intended to send an alarm to a supervisor should the user become incapacitated. In this scenario, the failure of the equipment (and lack of assistance from the colleague) results in the 'alarm' condition not being received or actioned by a 'supervisor' located adjacent to the hazardous area. This, in turn leads to fatality.

The scenario is modelled in Figure 13.2. Gate G1 models the demand placed on the safety-related system and Gate G2 models the mitigation. The events:

ATRISK	are the periods to which an individual is exposed
SEP	is the probability that the colleague is unavailable to assist
HE1	is the probability that the colleague fails to observe the problem
INCAP	is the probability that the colleague is incapacitated
DEMAND	is the probability that the incident arises during the event
FATAL	is the probability that the incident would lead to fatality if the worker is not rescued

Assume that the frequency of Gate G1 is shown to be 4.3×10^{-4} pa. Assume, also, that the target Maximum Tolerable Risk is 10^{-5} pa. In order for the frequency of the top event to equal 10^{-5} pa the probability of failure associated with Gate G2 must be $1 \times 10^{-5}/4.3 \times 10^{-4} = 2.33 \times 10^{-2}$. However, the event HE2 has been assigned a PFD of 10^{-2} which leaves the target PFD of the intercom to be $\mathbf{1.33 \times 10^{-2}}$.

Thus a **SIL 1 target (low demand)** will be placed on this safety function. Notice how critical the estimate of human error is in affecting the SIL target for the intercom. Had HE2 been 2×10^{-2} then the target PFD would have been $2.33 \times 10^{-2} - 2 \times 10^{-2} = 3.3 \times 10^{-3}$. In that case the target for the intercom would have **been SIL 2**.

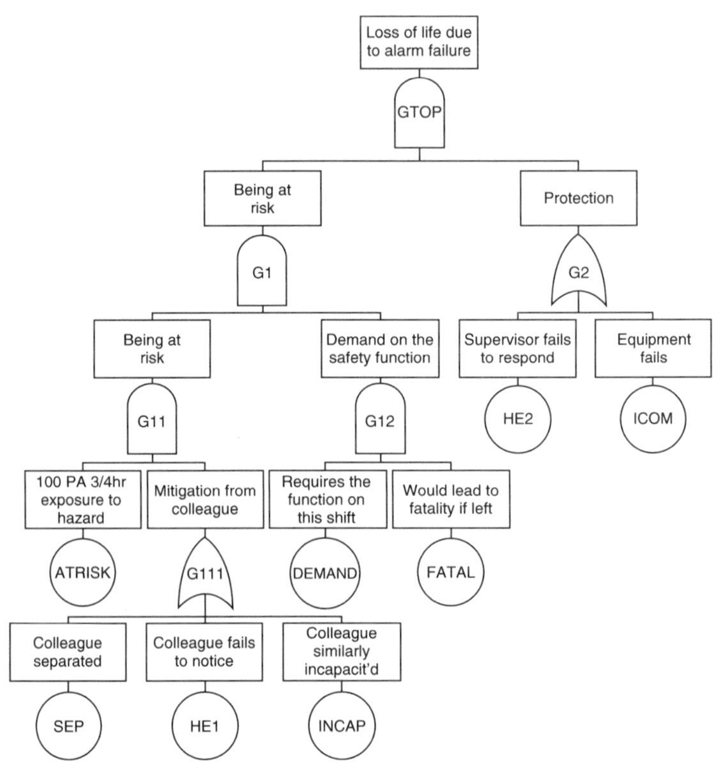

Figure 13.2
Loss of alarm function

13.3 Maximum tolerable failure rate involving alternative propagations to fatality

In this example, as a result of instrument and plant failures, a toxic gas cloud is released. Two types of hazard are associated with the scenario:

(a) CONCENTRATION OF GAS ON SITE
In this case a wind velocity of less than 1 m/sec is assumed as a result of which inversion would cause a concentration of gas within the site boundary, possibly leading to fatality.

Max Tolerable Risk $= 10^{-5}$ pa (perhaps 10^{-4} pa overall voluntary risk but 10 similar hazards)
Downstream pipe rupture due to 8 bar $= 10^{-2}$ pa
Wind <1 m/s assumed to be 1 day in 30 $= 3.3 \times 10^{-2}$

Plant in operation, thus causing exposure to the hazard, 100% of the time
Personnel close enough = 75%
Propagation of failure to fatality is estimated to be 80%
Thus **Max Tolerable PFD** = 10^{-5} pa/(0.01 pa × 3.3 × 10^{-2} × 0.75 × 0.8) = **5.1 × 10^{-2}**

(b) SPREAD OF GAS TO NEARBY HABITATION
In this case a wind velocity of greater than 1 m/sec is assumed and a direction between north and north west as a result of which the gas cloud will be directed at a significant area of population.

Max Tolerable Risk = 10^{-5} pa (public, involuntary risk)
Downstream pipe rupture due to 8 bar = 10^{-2} pa
Wind >1 m/s assumed to be 29 days in 30 = 97%
Wind direction from E to SE, 15%
Plant in operation, thus causing exposure to the hazard, 100% of the time
Public present = 100%
Propagation of failure to fatality is assumed to be 20%
Thus **Max Tolerable PFD** = 10^{-5} pa/(0.01 pa × 0.97 × 0.15 × 0.20) = **3.4 × 10^{-2}**

The lower of the two **Max Tolerable PFDs is 3.4 × 10^{-2}** which becomes the target.

SIL targets for the safety-related systems would be based on this. Thus, if only one level of protection were provided a **SIL 1 target** would apply.

13.4 Hot/cold water mixer integrity

In this example, a programmable equipment mixes 70°C water with cold water to provide an appropriate outlet to a bath. In this scenario, a disabled person is taking a bath, assisted by a carer. The equipment failure, which leads to the provision of 70°C water, is mitigated by human intervention.

Figure 13.3 models the events leading to fatality. Gate G11 apportions the incidents between those failures occurring prior to the bath (such that it is drawn with scalding water) (G111) and those that occur during the bath (G112). It was assumed

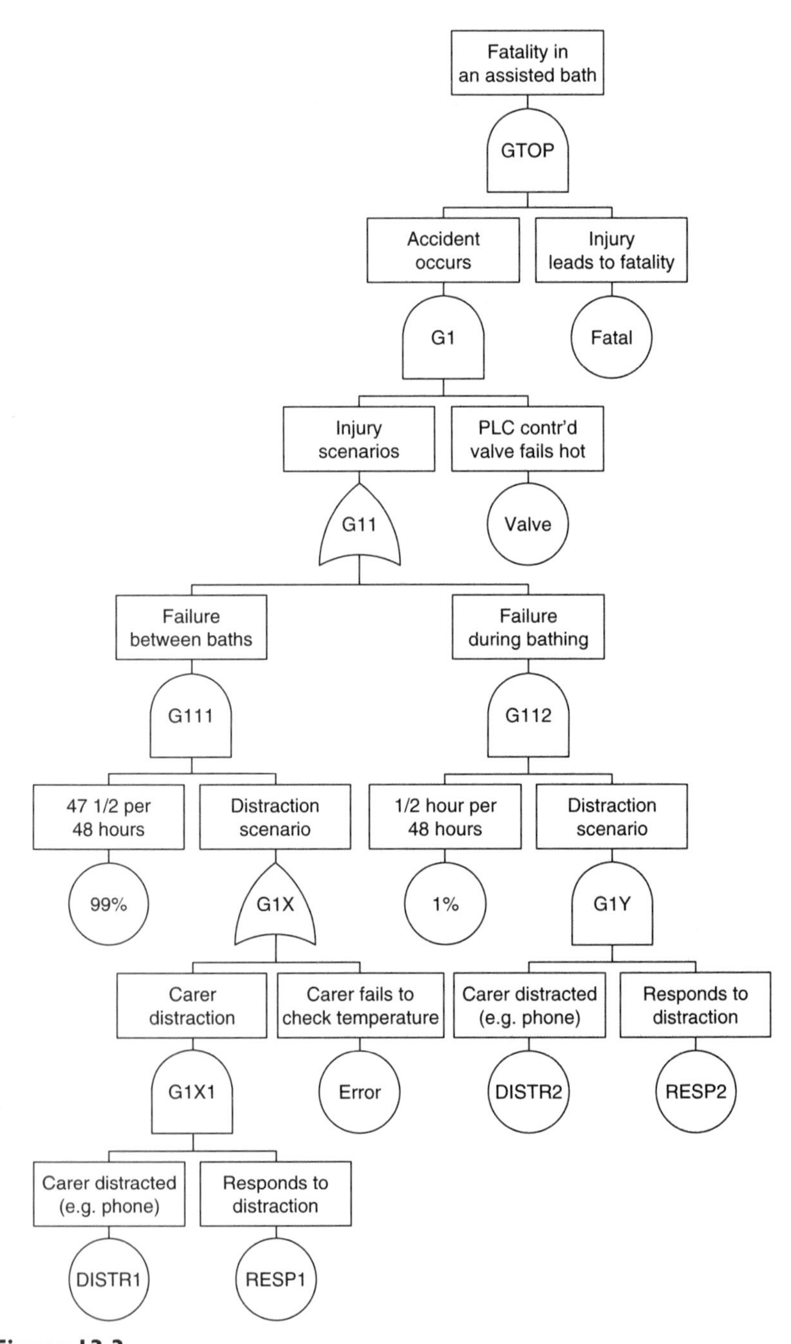

Figure 13.3
Fault tree – with assistance from a carer

that a bath occupies ½ hour per 2 days. Thus the probability of the former is 47½/48 = 99% and the latter therefore 1%.

A 20% chance of a distraction arising is assumed.
A 10% chance of the carer responding to the distraction is assumed.
The human error whereby the carer fails to detect a scalding bath is estimated as **0.1**.

The reader might care to study Figure 13.3 and verify that the probability associated with gate G11 is $(0.99 \times [0.1 \times 0.2 + 0.1]) + (0.01 \times [0.1 \times 0.2]) = \mathbf{0.119}$.

The probability of an incident becoming fatal has been estimated, elsewhere, as 8.1%. The maximum tolerable risk has been set as 10^{-5} pa, thus the maximum tolerable incident rate is $10^{-5}/8.1\% =$ **1.2×10^{-4} pa (Gate G1)**.

The maximum tolerable failure rate for the product is therefore:

$$\text{Gate G1/Gate G11} = 1.2 \times 10^{-4}\,\text{pa}/0.119 = \mathbf{1.01 \times 10^{-3}\,pa}.$$

This would imply a safety-integrity target of **SIL 2 (high demand)**.

13.5 Scenario involving high temperature gas to a vessel

In this example, gas is cooled before passing from a process to a vessel. The scenario involves loss of cooling which causes high temperature in the vessel, resulting in subsequent rupture and ignition. This might well be a three fatality scenario.

Supply profile permits the scenario (pilot alight)	100%
Probability that drum ruptures	5%
Probability of persons in vicinity of site (pessimistically)	50%
Probability of ignition	90%
Probability of fatality	100%

Assuming a maximum tolerable risk of 10^{-5} pa, the maximum tolerable failure rate is $10^{-5}\,\text{pa}/(0.05 \times 0.5 \times 0.9) =$ **4.4×10^{-4} pa**.

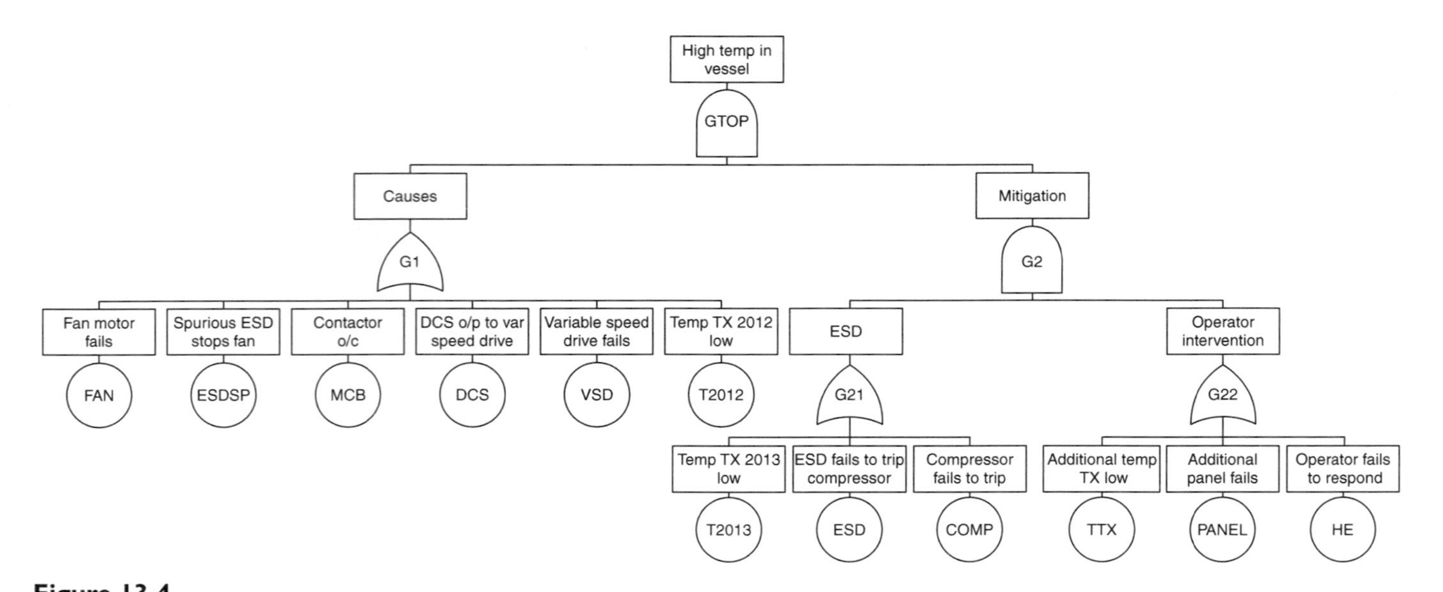

Figure 13.4
Fault tree – high temperature in vessel

The scenario is modelled in Figure 13.4. Only Gate G22 (involving human intervention and a totally independent equipment) is independent of the ESD (emergency shutdown system). If a probability of failure on demand in the SIL 1 range (say 3×10^{-2}) is assigned to Gate G22 then the top event target reduces to 4.4×10^{-4} pa/3×10^{-2} pa = 1.5×10^{-2} pa which is also SIL 1. Thus a **SIL 1 target (low demand)** is adequate for the ESD.

Assume that the frequency of the top event is **1.3×10^{-5} pa which meets the target**.

ALARP

If a cost per life saved criteria of £4 000 000 is used then the expenditure on any proposal which might reduce the risk to 10^{-7} pa (based on 10^{-6} pa but with 10 similar hazards) can be calculated (based on a 30-year plant life) as:

The frequency of the top event maps to a risk of $1 \times 10^{-5} \times (1.3 \times 10^{-5}/4.4 \times 10^{-4}) = 3 \times 10^{-7}$ pa and is thus in the ALARP region.

$$£4\,000\,000 = £\ \text{proposed}/([3 \times 10^{-7} - 1 \times 10^{-7}] \times 3\ \text{deaths} \times 30\ \text{yrs})$$

Thus £ proposed = £72

Any proposal involving less than £72, which would reduce the risk to 10^{-7} pa, should be considered. It is unlikely that any significant risk reduction can be achieved for that capital sum.

CHAPTER 14

HYPOTHETICAL RAIL TRAIN BRAKING SYSTEM (EXAMPLE)

The following example has been simplified and as a consequence some of the operating modes have been changed in order to maintain the overall philosophy but give clarity to the example.

14.1 The systems

In this example we have a combination of two safety-related systems. One is a 'high demand' train primary braking system, together with a second level of protection consisting of a 'low demand' emergency braking system.

Typically there are at least two methods of controlling the brakes on carriage wheels. The 'high demand' system would be the primary braking function activated by either the train driver or any automatic signalled input (such as ATP). This system would send electronic signals to operate the brakes on each bogie via an air operated valve. This is a proportional signal to regulate the degree of braking. The system is normally energised to hold brakes off. The output solenoid is de-energised to apply the brakes.

Each bogie has its own air supply reservoir topped up by an air generator. Air pressure has to be applied to operate the brakes. However, each bogie braking system is independent and each train has a minimum of two carriages. The loss of one bogie braking system would reduce braking by a maximum of 25%. It is assumed that the safety function is satisfied by three out of the four bogies operating (i.e. two must fail).

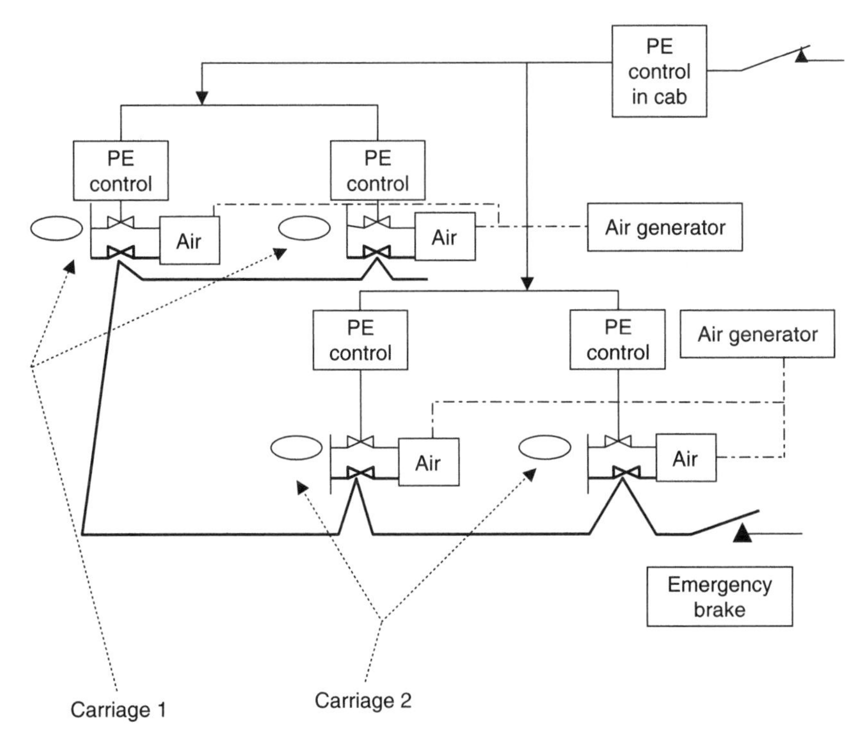

Figure 14.1
Braking arrangement

In addition to this primary braking system there is separate emergency braking. This is a single electrical wire loop that runs the full length of the train connected to an emergency button in the driver's cab. This circuit operates a normally energised solenoid valve. This circuit holds the brakes off and the emergency solenoids are de-energised to apply full braking pressure to the brakes.

Figure 14.1 shows the general arrangement of the two systems serving four bogies over two carriages.

14.2 The SIL targets

The specification for this design requires a **SIL 2 target for the primary braking system**, and a **SIL 3 target for the emergency braking system**.

These targets may have been arrived at by a risk graph approach. Therefore, unlike Chapter 11 where a specific quantified target was assessed, the SIL targets only provide an order of magnitude range of failure rates (or probabilities of failure on demand) for each of the two safety-related systems.

The SIL 2 braking system is a high demand system and, thus, the target is that the failure rate is **less than 10^{-2} pa**.

The SIL 3 emergency braking system is a low demand system and, thus, the target is that the probability of failure on demand is **less than 10^{-3}**.

It should be noted that the two systems are not independent in that they share the air power and brake actuator systems. As a result the overall safety integrity cannot be assessed as the combination of independent SIL 2 and SIL 3 systems. The common elements necessitate that the overall integrity is assessed as a combination of the two systems and this will be addressed in Section 14.6.

14.3 Assumptions

As in Chapter 11, assumptions are key to the validity of any reliability model and its quantification.

(a) Failure rates (symbol λ), for the purpose of this prediction, are assumed to be constant with time. Both early and wearout-related failures are assumed to be removed by burn-in and preventive replacement respectively.
(b) The majority of failures are revealed on the basis of 2 hourly usage. Thus, half the usage interval (1 hour) is used as the downtime.
(c) The proof-test interval of the emergency brake lever is 1 day. Thus the average downtime of a failure will be 12 hours.
(d) The common cause failure beta factor will be determined by the same method as in Chapter 11. A partial beta factor of 1% is assumed, for this example, in view of the very high inspection rate.
(e) The main braking cab PE controller operates via a digital output. The bogie PE operates the valve via an analogue output.

14.4 Failure rate data

Credible failure rate data for this example might be:

Item	*Failure mode*	*Failure rates (10^{-6} per hour) (total)*	*(mode)*	*MDT (hrs)*
PES (cab)	Serial output low	2	0.6	1
PES (bogie)	Analogue ouput low	2	0.6	1
Actuated valve	Fail to move	5	1.5	1
Solenoid valve	Fail to open	0.8	0.16	12
Driver's levers				
Emergency	Fail to open contact	1	0.1	12
Main	No braking	1	0.1	1
Bogie air reservoir	Fail	1	1	1
System (reservoir check valve and compressor) achieved by regular (daily use)				
Brake shoes	Fail	0.5	0.5	1
A low failure rate achieved by regular (2 weeks) inspection				
Common cause failure of air			0.05	
Common cause failure of brake shoes			0.005	

14.5 Reliability models

It is necessary to model the 'top event' failure for each of the two systems. Chapter 11 used the reliability block diagram method and, by contrast, this chapter will illustrate the fault tree approach.

14.5.1 Primary braking system (high demand)

Figure 14.2 is the fault tree for failure of the primary braking system. Gates G22 and G23 have been suppressed to simplify

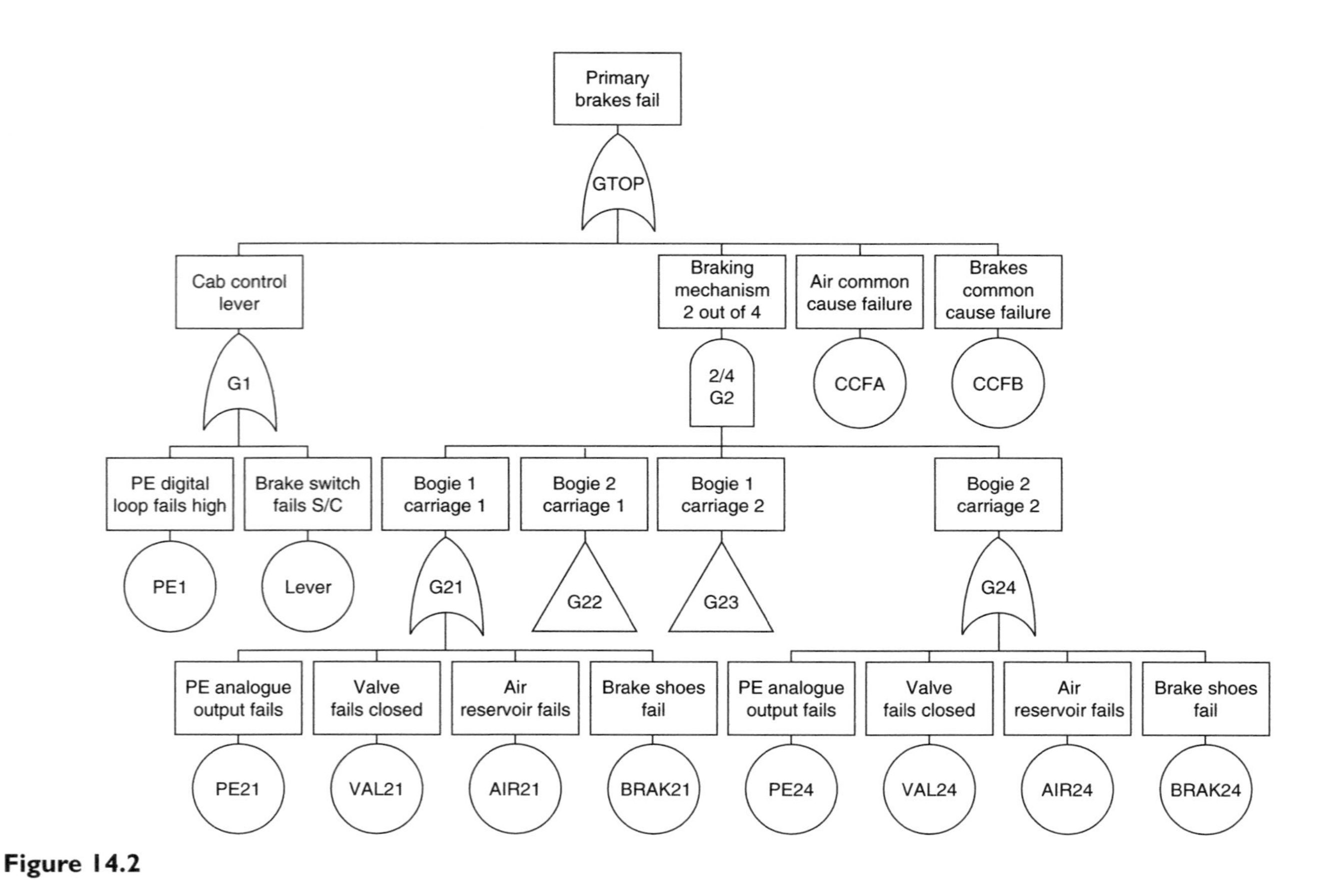

Figure 14.2
Fault tree for primary braking

Results of fault tree quantification for top event: GTOP

Top event frequency = 0.755E−06 per hour
0.662E−02 per year

Top event MTBF = 0.132E+07 hours
0.151E+03 years

Top event MDT = 0.100E+01 hours

Top event probability = 0.755E−06

Basic event reliability data

Basic event	*Type*	*Failure rate*	*Mean fault duration*	*Constant probability*
CCFA	I/E	.500E−07	1.00	
CCFB	I/E	.500E−08	1.00	
PE1	I/E	.600E−06	1.00	
LEVER	I/E	.100E−06	1.00	
PE21	I/E	.600E−06	1.00	
VAL21	I/E	.150E−05	1.00	
AIR21	I/E	.100E−05	1.00	
BRAK21	I/E	.500E−06	1.00	
PE22	I/E	.600E−06	1.00	
VAL22	I/E	.150E−05	1.00	
AIR22	I/E	.100E−05	1.00	
BRAK22	I/E	.500E−06	1.00	
PE23	I/E	.600E−06	1.00	
VAL23	I/E	.300E−05	1.00	
AIR23	I/E	.100E−05	1.00	
BRAK23	I/E	.500E−06	1.00	
PE24	I/E	.600E−06	1.00	
VAL24	I/E	.150E−05	1.00	
AIR24	I/E	.100E−05	1.00	
BRAK24	I/E	.500E−06	1.00	

Barlow-Proschan measure of cut set importance

Rank 1 Importance .795 MTBF hours .167E+07 MTBF years 190.

Basic event	*Type*	*Failure rate*	*Mean fault duration*	*Constant probability*
PE1	I/E	.600E−06	1.00	

Rank 2 Importance .132 MTBF hours .100E+08 MTBF years .114E+04

Basic event	*Type*	*Failure rate*	*Mean fault duration*	*Constant probability*
LEVER	I/E	.100E−06	1.00	

Rank 3 Importance .662E−01 MTBF hours .200E+08 MTBF years .228E+04

Basic event	*Type*	*Failure rate*	*Mean fault duration*	*Constant probability*
CCFA	I/E	.500E−07	1.00	

Rank 4 Importance .662E−02 MTBF hours .200E+09 MTBF years .228E+05

Basic event	*Type*	*Failure rate*	*Mean fault duration*	*Constant probability*
CCFB	I/E	.500E−08	1.00	

the graphics. They are identical, in function, to G21 and G24. Note that the Gate G2 shows a figure ‘2’, being the number of events needed to fail.

The frequency of the top event is **6.6×10^{-3} pa which meets the SIL 2 target**.

The table following Figure 14.2 shows part of the fault tree output from the Technis TTREE package (see end of book). The cut sets have been ranked in order of frequency since this is a high demand scenario which deals with a failure rate. Note that 80% of the contribution to the top event is from the PE1 event.

14.5.2 Emergency braking system (low demand)

Figure 14.3 is the fault tree for failure of the emergency braking system. Gates G22 and G23 have been suppressed in the same way as for Figure 14.2.

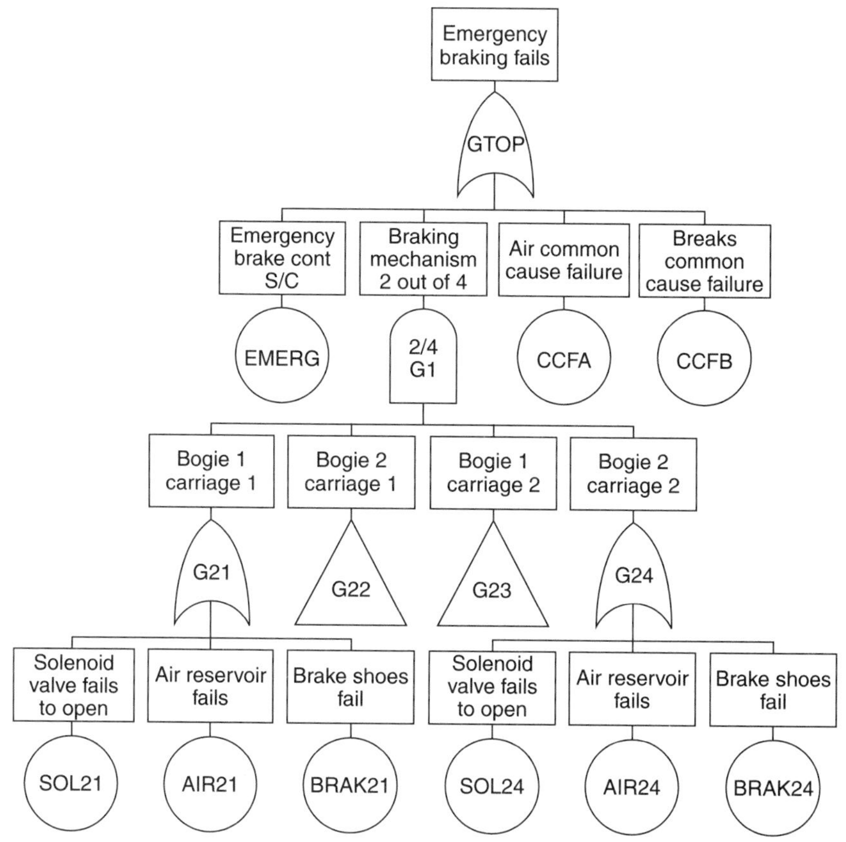

Figure 14.3
Fault tree for emergency braking

Results of fault tree quantification for top event: GTOP

Top event frequency	= 0.155E−06 per hour
	0.136E−02 per year
Top event MTBF	= 0.645E+07 hours
	0.736E+03 years
Top event MDT	= 0.809E+01 hours
Top event probability	**= 0.126E−05**

Basic event reliability data

Basic event	*Type*	*Failure rate*	*Mean fault duration*	*Constant probability*
EMERG	I/E	.100E−06	12.0	
CCFA	I/E	.500E−07	1.00	
CCFB	I/E	.500E−08	1.00	
SOL21	I/E	.160E−06	12.0	
AIR21	I/E	.100E−05	1.00	
BRAK21	I/E	.500E−06	1.00	
SOL22	I/E	.160E−06	12.0	
AIR22	I/E	.100E−06	1.00	
BRAK22	I/E	.500E−05	1.00	
SOL23	I/E	.160E−06	12.0	
AIR23	I/E	.100E−05	1.00	
BRAK23	I/E	.500E−06	1.00	
SOL24	I/E	.160E−06	12.0	
AIR24	I/E	.100E−05	1.00	
BRAK24	I/E	.500E−06	1.00	

Fussell-Vesely measure of cut set importance

Rank 1 Importance .956 Cut set probability .120E−05

Basic event	*Type*	*Failure rate*	*Mean fault duration*	*Constant probability*
EMERG	I/E	.100E−06	12.0	

Rank 2 Importance .398E−01 Cut set probability .500E−07

Basic event	*Type*	*Failure rate*	*Mean fault duration*	*Constant probability*
CCFA	I/E	.500E−07	1.00	

Rank 3 Importance .398E−02 Cut set probability .500E−08

Basic event	*Type*	*Failure rate*	*Mean fault duration*	*Constant probability*
CCFB	I/E	.500E−08	1.00	

Rank 4 Importance .765E−05 Cut set probability .960E−11

Basic event	*Type*	*Failure rate*	*Mean fault duration*	*Constant probability*
SOL21	I/E	.160E−06	12.0	
BRAK22	I/E	.500E−05	1.00	

The probability of the top event is **1.3×10^{-6} which meets the SIL 3 target with approximately 2 orders of magnitude margin**.

The table following Figure 14.3 shows part of the fault tree output as in the previous section. In this case the cut sets have been ranked in order of probability since this is a low demand scenario which deals with a PFD. Note that >95% of the contribution to the top event is from the EMERG event (lever).

14.6 Overall safety-integrity

As mentioned in Section 14.2 the two safety-related systems are not independent. Therefore the overall failure rate (made up of the failure rate of the primary braking and the PFD of the emergency braking) is calculated as follows. The fault tree in Figure 14.4 combines the systems and thus takes account of the common elements in its quantification.

The overall failure rate is **4.8×10^{-4} pa**. The cut set rankings show that the air supply Common Cause Failure accounts for 90% of the failures.

This example emphasises that, since the two systems are not independent, one cannot multiply the failure rate of the primary braking system (6.6×10^{-3} pa) by the PFD of the emergency braking system (3.6×10^{-6}). The result would be nearly 4 orders optimistic and the overall arrangement has to be modelled as shown in Figure 14.4.

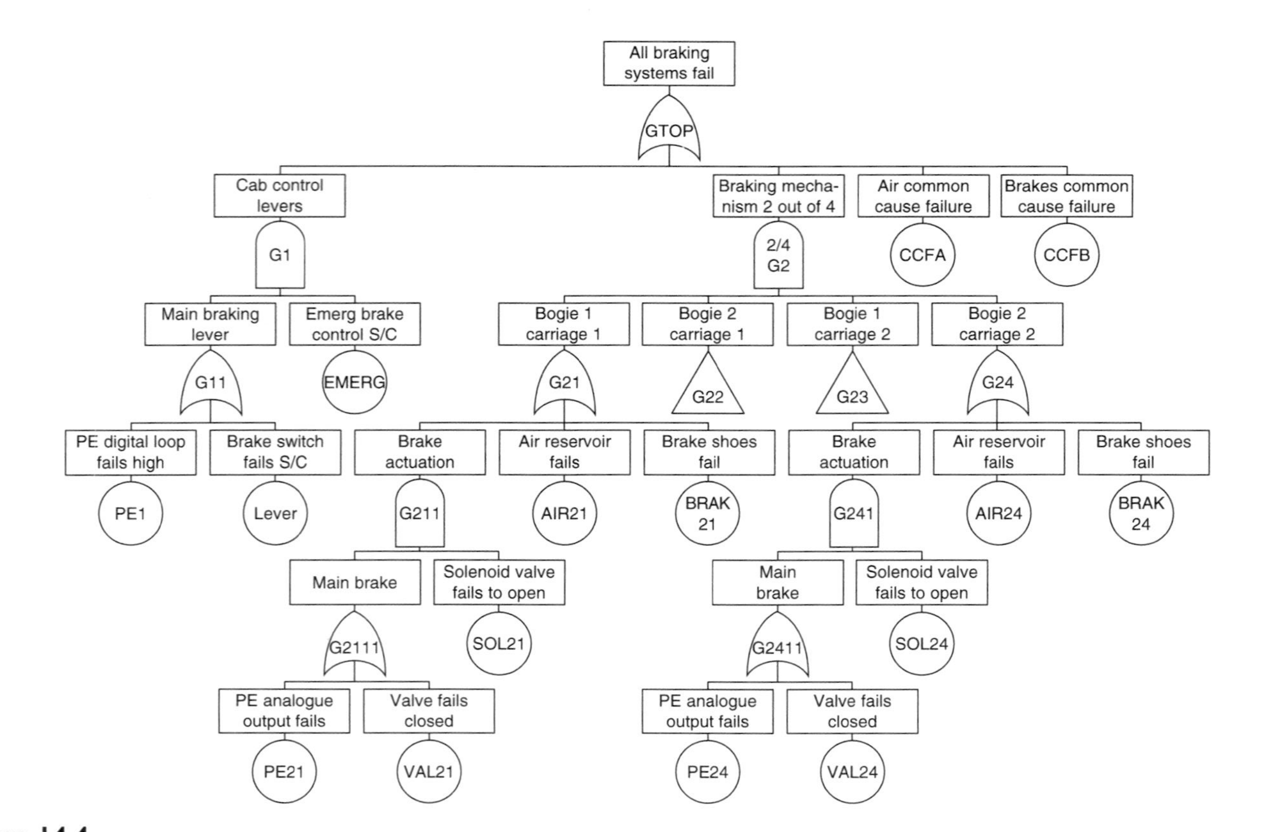

Figure 14.4
Combined fault tree

Results of fault tree quantification for top event: GTOP

Top event frequency = 0.550E−07 per hour
0.482E−03 per year

Top event MTBF = 0.182E+08 hours
0.207E+04 years

Top event MDT = 0.100E+01 hours

Top event probability = 0.550E−07

Basic event reliability data

Basic event	*Type*	*Failure rate*	*Mean fault duration*	*Constant probability*
CCFA	I/E	.500E−07	1.00	
CCFB	I/E	.500E−08	1.00	
EMERG	I/E	.100E−06	12.0	
PE1	I/E	.600E−06	1.00	
LEVER	I/E	.100E−06	1.00	
AIR21	I/E	.100E−05	1.00	
BRAK21	I/E	.500E−06	1.00	
SOL21	I/E	.160E−06	12.0	
PE21	I/E	.600E−06	1.00	
VAL21	I/E	.150E−05	1.00	
AIR22	I/E	.100E−05	1.00	
BRAK22	I/E	.500E−06	1.00	
SOL22	I/E	.160E−06	12.0	
PE22	I/E	.600E−06	1.00	
VAL22	I/E	.150E−05	1.00	
AIR23	I/E	.100E−05	1.00	
BRAK23	I/E	.500E−06	1.00	
SOL23	I/E	.160E−06	12.0	
PE23	I/E	.600E−06	1.00	
VAL23	I/E	.300E−05	1.00	
AIR24	I/E	.100E−05	1.00	
BRAK24	I/E	.500E−06	1.00	
SOL24	I/E	.160E−06	12.0	
PE24	I/E	.600E−06	1.00	
VAL24	I/E	.150E−05	1.00	

Barlow-Proschan measure of cut set importance

Rank 1 Importance .909 MTBF hours .200E+08 MTBF years .228E+04

Basic event	*Type*	*Failure rate*	*Mean fault duration*	*Constant probability*
CCFA	I/E	.500E−07	1.00	

Rank 2 Importance .909E−01 MTBF hours .200E+09 MTBF years .228E+05

Basic event	*Type*	*Failure rate*	*Mean fault duration*	*Constant probability*
CCFB	I/E	.500E−08	1.00	

Rank 3 Importance .363E−04 MTBF hours .500E+12 MTBF years .571E+08

Basic event	*Type*	*Failure rate*	*Mean fault duration*	*Constant probability*
AIR21	I/E	.100E−05	1.00	
AIR22	I/E	.100E−05	1.00	

APPENDIX I

FUNCTIONAL SAFETY CAPABILITY – TEMPLATE PROCEDURE

This procedure could be part of a company's Quality Management System (e.g. ISO 9001). It contains those additional practices (over and above ISO 9001) necessary to demonstrate Functional Safety Capability as would be assessed in a CASS type 5 assessment (see Chapter 10).

A large organisation, with numerous activities and product lines, might require more than one procedure whereas a small company would probably find a single procedure satisfactory.

This template has been successfully used by a medium to large sized company in the safety systems integration field. It consists of one (this) main procedure and five work practices to cover details of safety assessment (see Annex 1).

Terms (e.g. Safety Authority, Technical Authority) are examples only, and will vary from organisation. xxxs are used to designate references to company procedures.

COMPANY STANDARD xxx IMPLEMENTATION OF FUNCTIONAL SAFETY

CONTENTS

1. Purpose of document

This standard provides detail of those activities related to setting and achieving specific safety-integrity targets and involves the design, installation, maintenance and modification stages of the life-cycle. Where the activity in question is already catered for elsewhere in the XYZ Ltd quality management system, this document will provide the appropriate reference.

2. Scope

The standard shall apply to all products and documentation designed, produced, installed or supported by XYZ Ltd except where contract requirements specifically call for an alternative.

3. Terms and abbreviations

As appropriate……

4. Applicable documents and references

4.1 ISO 9000-2000 Quality Systems Model for Design, Development, Manufacture and Service.

4.2 IEC 61508 Functional Safety of Electrical/Electronic/Programmable Electronic Safety-related Systems.

4.3 Safety, Competency and Commitment (Competency Guidelines for Safety-related Practitioners) ISBN 0 85296 787X.

4.4 Procedures:
XYZ Ltd Procedure FS/PROC/001: RAMS Quantification
XYZ Ltd Procedure FS/PROC/002: Hazard Identification
XYZ Ltd Procedure FS/PROC/003: Safety Integrity
XYZ Ltd Procedure FS/PROC/004: Failure Rate Data
XYZ Ltd Procedure FS/PROC/005: Demonstration of SIL Compliance

The structure of the XYZ Ltd standards and procedures for functional safety is shown in Appendix 1.

5. Functional safety policy

Paragraph x of the Quality Manual (QMxxx) emphasises that capability in respect of functional safety is a specific design capability within XYZ Ltd. Some contracts will relate to safety-related applications. Some developments will specifically target safety-integrity conformance as a design requirement.

For these instances the provisions of International Standard IEC 61508 (and related guidance) shall be met by the XYZ Ltd quality management system.

6. Competencies

HR department will maintain a 'safety-related competence register' containing profiles of those individuals eligible to carry out functional safety assessment and design tasks. Periodically the Managing Director and Safety Engineering Manager will review the list.

The list will be updated from:

- Individuals' attendance at relevant off-the-job courses
- Records of SR experience from each project (on-the-job training) (Project Managers will provide this information to the Personnel Manager)
- Details of new employees or contractors

The register will be based on (although not restricted to) the 12 basic SR tasks listed in Appendix A of reference 4.3. Guidance on the competencies needed for each task is given in reference 4.3 although XYZ Ltd will continue to develop its interpretation of that guidance in the light of experience.

SAMPLE ENTRY IN THE COMPETENCY REGISTER

Competency Assessment
Mr A N OTHER
Technical Manager
d.o.b. *31.2.1950*
Quals. BSc, MSaRS, MSc (Safety and Reliability Heriot Watt)
d.o. employment by XYZ Ltd *1.3.1920*

Application domain knowledge
- *Sectors – oil and gas*
- *Project xxx – SR code in 'C' (SIL 2 – 1 year's experience)*

Accuracy and detail
- *Project xxx*

Decisions/Communication/Inter-working
- *Good (see records of 2.1.01 appraisal)*

FS assurance
- *NO*

Functional safety and regulatory knowledge
- *Attended in-house course 31.2.88*
- *Knows 61508; has reviewed this procedure with external consultant*
- *Participated in FMEA of SFF (2223)*

Testing
- *Participated in xxx*

Reviews
- *NO*

FS audits
- *Reviewed the xxxx Audit (2.1.99)*

Bidding for work
- *NO*

Safety authority
- *Project xxx 1999 (SIL 2)*

Assessing individuals on this register
- *NO*

Example of specific jobs involving SR competencies include:

Safety Authority

Each project has a Safety Authority who is independent of the project activities themselves. A Safety Authority will have had previous experience of a project involving similar hardware and the same language in a similar application. He/she will have received the XYZ Ltd *training course on Functional Safety. He/she will have had experience of at least one Safety-Integrity Assessment.*

Functional Safety Auditor

Functional Safety Audits (section 9) are carried out by a person other than the Safety Authority for a project. He/she will have received the XYZ Ltd *training course on Functional Safety. He/she will have had experience of at least one Safety-Integrity Assessment.*

Safety Engineering Manager
The SEM will provide the company's central expertise in functional safety. He/she will have substantial experience in functional safety assessment and will be thoroughly conversant with IEC 61508 and related standards.

For each project, the Project Manager (assisted by the Safety Authority) shall consult the competence register to decide who will be allocated to each task. In the event that a particular competence(s) is not available then he will discuss the possible options involving training, recruitment or subcontracting the task with the Managing Director.

Each individual on the competency register will participate in an annual review (generally at the annual appraisal) with his/ her next level of supervision competent to assess this feature of performance. He/she will also discuss his/her recent training and experience, training needs, aspirations for future SR work.

7. Safety-related activities

7.1 Contract or project review

Where a bid, or invitation to tender, explicitly indicates an SR requirement (e.g. reference to IEC 61508, use of the term safety-critical etc.) then the Sales Engineer will consult a Safety Authority for advice.

All contracts (prior to acceptance by XYZ Ltd) will be examined to ascertain if they involve safety-related requirements. These requirements may be stated directly by the client or may be implicit by reference to some standard. Clients may not always use appropriate terms to refer to safety-related applications or integrity requirements. Therefore, the assistance of the Safety Engineering Manager will be sought before a contract is declared not safety-related.

A project or contract may result in there being a specific integrity requirement placed on the design (e.g. SIL 2 of IEC

61508). Alternatively, XYZ Ltd may be required to advise on the appropriate integrity target in which case FS/PROC/003 will be used.

7.2 Assigning responsibilities

For each PROJECT or contract the Project Manager shall be responsible for ensuring (using the expertise of the Safety Authority) that the safety-integrity requirements are ascertained and implemented.

Each project will have a Technical Authority and a Safety Authority (see section 6 of this standard).

The Project Manager will ensure that the FS activities (for which he carries overall responsibility to ensure that they are carried out) called for in this standard (and related procedures) are included in the project Quality and Safety Plan and the life-cycle techniques and measures document. Specific allocation of individuals to tasks will be included in the Quality and Safety Plan. These shall include:

- Design and implementation tasks (section 7.4(a) of this standard)
- Functional safety assessment tasks (section 7.4(b) of this standard)
- Functional safety audits (section 10 of this standard)

The Project Manager will ensure that the tasks are allocated to individuals with appropriate competence. The choice of individual may be governed by the degree of independence required, for an activity, as addressed in section 10 of this standard.

7.3 Quality and Safety Plan and life-cycle activities

Every project shall involve a Quality and Safety Plan which is the responsibility of the Project Manager. It will indicate the safety-related activities, the deliverables (e.g. safety-integrity assessment report) and the competent persons to be used. The Project Manager will consult the competency register and will review the choice of personnel with the Safety Authority.

The tasks are summarised in section 7.4 of this standard. Minimum SR items required in the Quality and Safety Plan are shown in Annex 2.

7.4 Assessment and design techniques

The life-cycle activities are summarised in this section and are cross-referenced to IEC 61508. They are implemented, by XYZ Ltd, by means of the Quality Management System (to ISO 9001 standard) by means of this standard and the associated Functional Safety Procedures (FS/PROCs/001-005) – see Annex 1.

7.4(a) Design and implementation tasks (referring to IEC 61508 tables and para. numbers)

1. *Organising and managing the design:* This is achieved by means of this, and related, procedures. This will satisfy 'Project management' as referred to in Tables B1, B2, B3, B4 of Part 2.
2. *Safety-related requirements in the specification:* This includes understanding the EUC boundary and its safety requirements and the scope of hazards and risks. It also includes the need for adequate clear documentation, appropriate reference to checklists, semi-formal or formal methods, CAD tools and separation of functions and of redundant hardware. It is also necessary to specify the SR functions and SILs allocated to sub-units and to structure the development into life-cycle stages. This references to Table B2 of Part 2 in respect of the hardware and Table A1 of Part 3 in respect of software.

- Safety functions should be described using semi-formal methods, an example is the use of cause and effect diagrams to describe control and shutdown logic. At SIL 3 semi-formal methods are required for all parts of the specification. At SIL 4 formal methods (e.g. VDM) are required for the SR elements.
- System response times, self-test functions, serial link error detection techniques, operator/maintainer interfaces should be included.

3. *Carrying out the design/development:* Tables B2 and A16–A18 of Part 2 and Tables A2–A5 of Part 3 contain the techniques relevant to design.

These include the need to evidence a structured design for both hardware and software and in particular a detailed life-cycle model for the software development. Issues of modularisation, proven components, methodologies and checklists are included in this heading (Table B2 Part 2 and A2 Part 3).

Software design methods (e.g. semi-formal methods, coding standards, defensive programming, modular programming) are covered by Table A4 of Part 3. In particular:

- Off-the-shelf hardware should have 10 device years' experience for SILs 1/2 and 20 device years for SILs 3/4.
- EUC and SR functions should be functionally separate in all cases. At SIL 3 there should be physical separation and at SIL 4 complete data/software separation together with some aspect of third party certification.
- Manageable module size, with single entry/exit, should be a conscious decision for all SILs. At SIL 3 and above methodologies (e.g. Yourdan) should be considered. At SIL 3 and above there must be conscious limitation of interrupts, no dynamic variables and a high level of on-line checking.

Fault tolerance techniques are listed in Tables A16–A17 Part 2 and Table A2 Part 3. An appropriate combination will be chosen for the application and the SIL.

- At SIL 1/2 a minimum of a CPU watchdog is required. At SIL 3 memory and I/O checks should employed and at SIL 4 a significantly 'state-of-the-art' combination of techniques.

4. *Architectural design constraints:* Ensure that the minimum architectural requirements are satisfied and take appropriate action vis-à-vis safe fail fraction and diagnostic coverage to achieve this. Tables 2 and 3 of Part 2 apply.

5. *Design to environmental requirements:* Table A17 of Part 2 applies and addresses such items as voltage-related parameters, separation, temperature changes and use of diversity.

- At SIL 3 there should be significant separation of multiple data lines and at SIL 4 total separation.
- Hardware should be either tested or certified to an appropriate standard for the application.

6. *Support tools:* Libraries of modules, translators, language subsets etc. are covered by Table A3 of Part 3.

- An applications suitable language is a requirement for all SILs. A coding standard must be used for SIL 2 upwards.

For SIL 3/4 certified tools or proven experience and a defined language sub-set are needed.

7. *Verification: Design review and test strategy to embrace SR features:* This includes design reviews, system integration and all types of test. Integration is covered by tests listed in Table B3 of Part 2 and Tables A5 and A6 of Part 3. Types of testing are covered by Table B5 (wrongly labelled 'validation') of Part 2 and Table A9 of Part 3, and Table A7 (wrongly labelled 'validation') of Part 3.

- Tests should include specified outputs for combinations of inputs, responses to unspecified inputs and use of boundary values. SILs 3/4 should include specific test cases for all critical logic elements.
- Module testing should include code inspections.

8. *Modifications:* Configuration control and testing the effect of modifications are covered by Table A8 of Part 3.

- There must be an impact analysis of all changes. At SIL 2 and above affected modules must be reverified and at SIL 3 and above the whole system.

9. *Producing and implementing a validation plan:* This should be included in the Quality and Safety Plan (see section 11 of this standard).

- The validation plan should embrace all the requirements and procedures. The aim is to ensure that all tests, audits, assessments and reviews are closed out with all the remedial actions completed.

10. *Planning and implementing functional safety in operations and maintenance:* **This item is only relevant where** XYZ Ltd **undertakes a support contract.** This may be included in the Quality and Safety Plan. Specific items (e.g. limited operational functions, user friendliness, clear instructions) are covered in Table B4 of Part 2 and Table A18 of Part 3.

- Items to be included are concise procedures, records of proof tests, records of demands on low demand systems, password protection, control of site modifications, training, evidence that human interface factors have been addressed.

11. *Planning installation and commissioning:* This may be included in the Quality and Safety Plan.
12. *Sector specific guidance:* Where sector specific guidance (e.g. IEC 61511, IGEM SR/15, UKOOA) is called for this will be identified in the Quality and Safety Plan. The Safety Authority will ensure that the requirements of the sector specific guidance and IEC 61508 are harmonised.
13. *Acquired modules (hardware and/or software):* These must involve verification of their random hardware failure reliability, proof-test methods, fault tolerant features and any statistical evidence of an achieved SIL claim. Documentary evidence is needed.

7.4(b) Functional safety targeting and assessment tasks

The Project Manager (with the advice of the Safety Authority) shall define, in the draft and subsequent issues of the Quality and Safety Plan, the points in the life-cycle where safety assessments will be carried out and the personnel who will conduct them. In general these will be undertaken by the Safety Authority. This is covered by Table A10 of Part 3 and Paragraph 7.4.3 of Part 2.

1. *Hazard identification (FS/PROC/002)*
2. *Integrity targeting (FS/PROC/003)*
3. *Assessment of random hardware failures (FS/PROC/001)*
4. *Assessment of safe-fail fraction (FS/PROC/001)*
5. *Assessment of conformance to qualitative features (FS/PROC/005)*

7.5 Method of documentation

This is dealt with in XYZ Standard STD/xxx. Specific documents (e.g. SIL assessment, FS audit) required for functional safety shall be called for in the Quality and Safety Plan as required.

8. Design and implementation

8.1 Corrective action and follow-up

During design, test and build, defects are recorded on 'Defect Reports'. During site installation and operations they are recorded on 'Incident Reports' which embrace a wider range of incident.

Problems elicited during design review will be recorded on form xxxx. Failures during test will be recorded as indicated in STD/xxx (factory) and PROC/xxx (site).

All defect reports will be copied to the Technical Authority who will decide if they are SR or not SR. He will positively indicate SR or not SR on each report. All SR reports will be copied to the Safety Authority who will be responsible for following up and closing out remedial action.

All SR incident reports, defect reports and records of SR system demands will be copied to the XYZ Ltd Safety Engineering Manager who will maintain a register of failures/ incidents. A 6 monthly summary (identifying trends where applicable) will be prepared and circulated to Project Managers and Technical Authorities and Safety Authorities.

8.2 Hazardous incidents

This is dealt with in QA/PROCxxx para x.

Incidents during site operations will be dealt with by the Project Manager. In the first instance, SR incidents subsequent to a project will be dealt with by the XYZ Ltd Managing Director who will instigate appropriate action. In both cases the cause of the incident will be studied in order to determine if it has an impact on the safety integrity of any other past or current contract. Appropriate remedial action (*vis-à-vis* design or maintenance) will be implemented.

It is necessary to maintain information on potential hazards (discovered as a result of assessments) and actual hazards (discovered from applications). CSD/xxx and xxx deal with this requirement and create the need for a central register. Project Managers will consult this register on a regular basis since all new entries will be circulated to all Project Managers.

8.3 Modifications and configuration control

This is dealt with in STD/xxx. Change proposals will be positively identified, by the Safety Authority, as SR or not SR. All SR change proposals will involve a design review before approval.

8.4 Operations and maintenance

This is dealt with in on-site work instructions CSD/PROC/xxx.

Where XYZ Ltd is responsible for operating or maintaining a system, Site Incident Reporting forms will be used. Any equipment failure or incident, perceived to have SR implications, will be recorded, by the XYZ Ltd Site Engineer.

The Project Manager shall monitor all failure report forms for safety-related implications and take remedial action as necessary.

The client will be encouraged to report (to XYZ Ltd) all demands on a low demand SR system by making the data recording a feature of the Operating Manual. In the case of an XYZ Ltd support contract the Project Manager will report demands to the Safety Engineering Manager.

8.5 Vendors and subcontractors

The Project Manager is responsible for ensuring that safety-integrity level targets are reflected in the requirements for purchased equipment, software and subcontract effort.

The Safety Authority shall review all vendor questionnaires and follow up claims for functional safety capability in order to establish how this is demonstrated.

This will also apply to the competency claims of subcontractors and consultants carrying out functional safety-related tasks.

9. Functional safety audits

The Project Manager (in consultation with the Safety Authority) shall define, in the Quality and Safety Plan, the points in the life-cycle where safety audits (minimum of one per project) will be carried out and the personnel who will conduct them.

The audit will be led by a competent person (see section 6 and the competency register) and the independence requirements of section 10 will be addressed. This document (and associated procedures) will be used as the basis for determining conformance. The auditor will prepare an audit plan prior to the audit. Previous audits will be consulted in order to ensure a representative coverage of auditable features. An audit report will be produced by the audit leader and endorsed by the Project Manager. Any unresolved difference of opinion will be reviewed by the Managing Director. Remedial actions will state a time scale after which each will be reviewed. A report will be issued and reviewed until the remedial actions have been resolved. FS audits will also meet the provisions of STD/xxx (audit).

10. Independence

In accordance with IEC 61508, the following degree of independence will be applied to assessments.

10(a) Establishing the SIL target

The level of independence to be applied when carrying out this process is recommended, according to consequence, as:

Multiple fatality say >5	Independent organisation
Multiple fatality	Independent department
Single fatality	Independent person
Injury	Independent person

For scenarios involving fatality, add one level of independence if there is lack of experience, unusual complexity or novelty of design.

10(b) Assessing conformance to the SIL target

SIL	Assessed by:
4	Independent organisation
3	Independent department
2	Independent person
1	Independent person

For SILs 2 and 3 add one level of independence if there is lack of experience, unusual complexity or novelty of design.

11. Validation

The validation, which will be called for in the Quality and Safety Plan and is specified in section 7.4(a) of this standard, will involve a Validation Plan. This plan will be prepared by the Safety Authority and will consist of a list of all the SR activities for the project, as detailed in this standard and related procedures.

The SA will produce a Validation Report which will remain active until all remedial actions have been satisfied. The SA, TA and Project Manager will eventually sign off the report which will form part of the Project File.

Annex 1

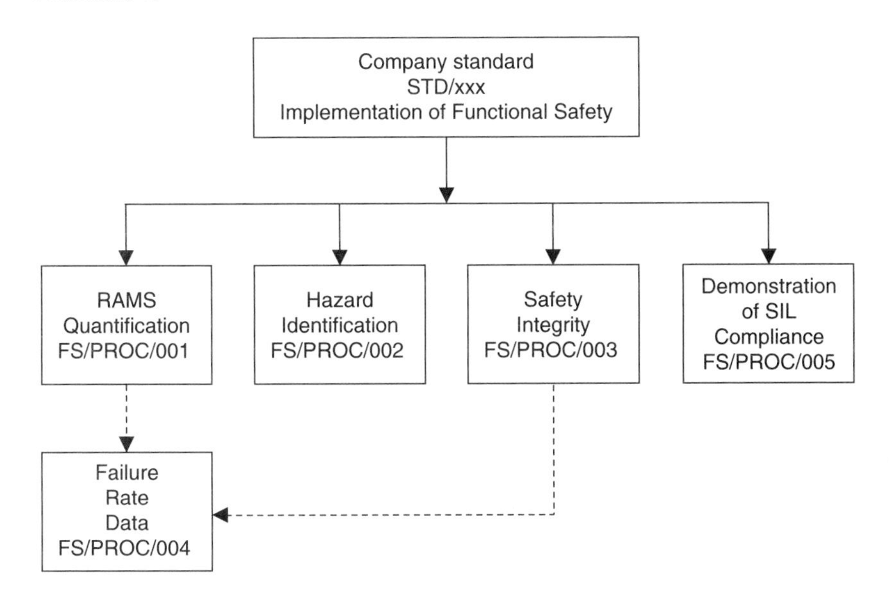

Annex 2 – Items for inclusion in the Quality and Safety Plan

Responsibilities:
Project Manager
FS authority for the project
FS assessment
FS audit
Validation

Items to be called for and described in outline:

Document hierarchy	(e.g. Requ's Spec, Software Spec, H/W drawings, code listings, Test Plan and results, Review Plan and results, Validation Report)

List of hardware modules

Software life-cycle	(e.g. S/W spec, flow charts, listing)
Tools	(e.g. PC and Visual Basic compiler ......)
Review Plan	(e.g. design review of Functional Spec and of code listings)
Test Plan	(e.g. list of module tests, functional test, acceptance test – including SR tests related to the SR failures described above)
Validation Report	(i.e. could be in the form of a matrix of requirements cross referenced to test results etc.)

Descriptions of:

The boundary of the SR system	(e.g. input signals, x IMPs, y Inverters etc.)
Description of failure modes	(e.g. spurious move, spurious release etc.)
SIL targets	(e.g. SIL 2 for all SR functions)

Annex 3 – Process flow diagrams

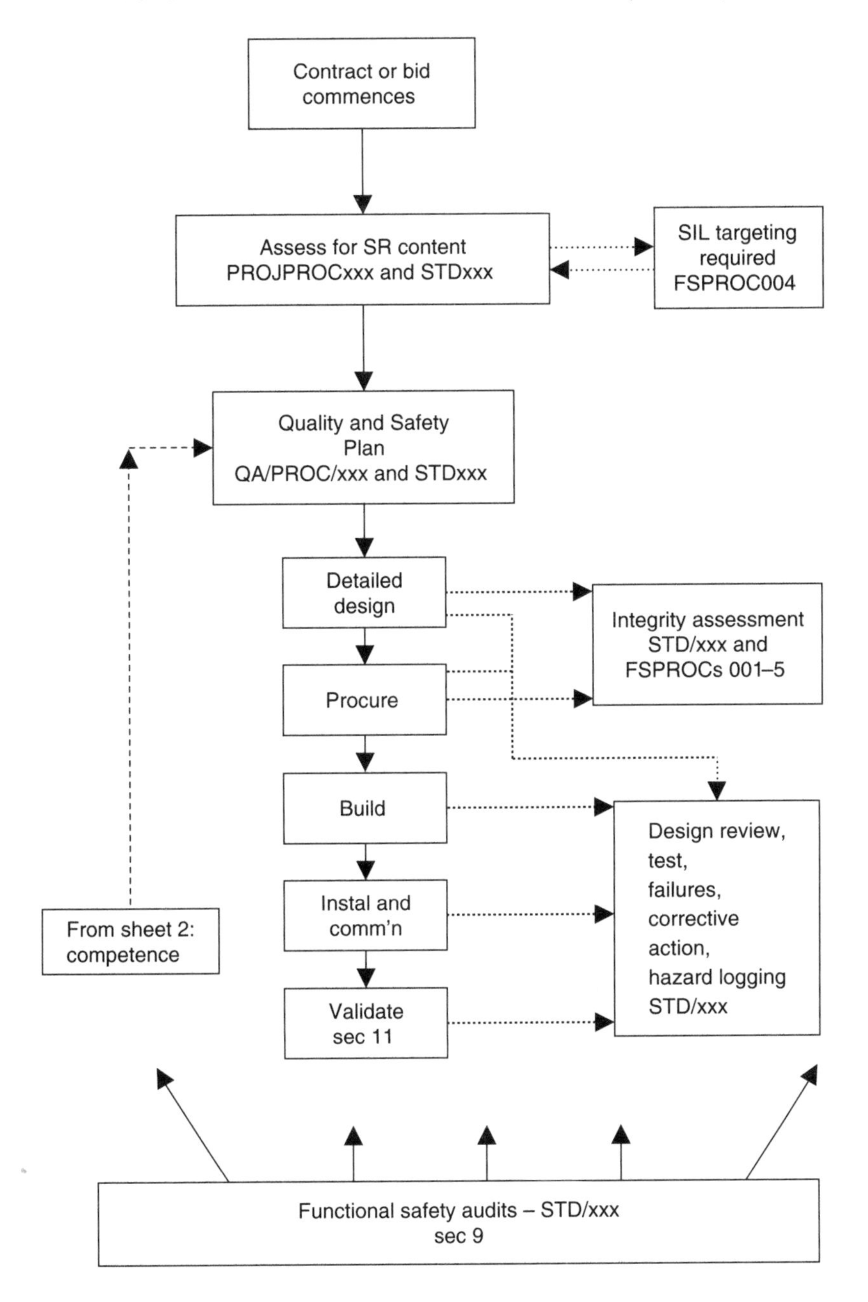

FUNCTIONAL SAFETY PROCESS FLOW DIAGRAM (SHEET 2)

To sheet 1

Competence register

Name: A N Other

SR Quals: e.g. MSaRS

SR TRg: xxxxx
Xxxxx
Xxxxx

SR Project Work
Xxxxxx
Xxxxxx

Recruitment

Training

Project work

Periodic review
STD/xxx

A NOTE ON THE SECOND LEVEL PROCEDURES 001–005

Procedure FS/PROC/001: RAMS Quantification
Will describe techniques to be used (see Chapter 6 of this book).

Procedure FS/PROC/002: Hazard Identification
Will deal with how the hazardous failures are identified.

Procedure FS/PROC/003: Safety Integrity
Will describe the approach to SIL targeting (see Chapter 2 of this book).

Procedure FS/PROC/004: Failure Rate Data
Will list sources of failure rate data to be used (see Chapter 7 of this book).

Procedure FS/PROC/005: Demonstration of SIL Compliance
This would closely resemble the demonstration templates given in Chapters 3, 4 and 5 of this book.

Appendix 2

Assessment schedule (checklist)

The following checklist assists in providing CONSISTENCY and RIGOUR when carrying out an Integrity Assessment. The checklist can be used to ensure that each of the actions have been addressed. Furthermore it can be included, as an Appendix, in an assessment report with the Paragraph Numbers of the report referenced against each item. In this way a formal review of rigour can be included.

1. Defining the assessment and the safety system

1.1 Describe the reason for the assessment, for example safety case support, internal policy, contractual requirement for IEC 61508. Paragraph No ......

1.2 Confirm the degree of independence called for and the competence of the assessor. This includes external consultants. Paragraph No ......

1.3 Define the safety-related system. This may be a dedicated item of safety-related equipment (i.e. ESD) or a control equipment which contains safety-related functions. Paragraph No ......

1.4 Define the various parts/modules of the system being studied and list the responsibilities for design and maintenance. For example, the PLC may be a proprietary item which has been applications programmed by the supplier/user – in which case information will be needed from the supplier/user to complete the assessment. Paragraph No ......

1.5 Describe the customer for, and deliverables anticipated for, the assessment. For example, 'XYZ to receive draft and final reports'. Paragraph No ......

1.6 Provide a justification, for example that the SIL calculation yields a target of less than SIL 1, where it is claimed that an equipment is not safety related. Paragraph No ......

1.7 Establish that the development (and safety) life-cycle has been defined for the safety-related system. Paragraph No ......

1.8 Establish that the Quality Plan (or other document) defines all the necessary activities for realising the requirements of IEC 61508 and that all the necessary design, validation etc. documents are defined.

2. Describing the hazardous failure mode and safety targets

2.1 Establish the failure mode(s) which is addressed by the study, against which the safety-related system is deemed to be a level of protection (for example, downstream overpressure for which ESD operates a slamshut valve). Paragraph No ......

2.2 Establish the risk criteria for the failure mode in question. Paragraph No ......

2.3 Taking account of the maximum tolerable risk, calculate the SIL(s) for the safety-related system for the failure mode(s) in question. Indicate whether the SIL has been calculated from a risk target, for example Table 2.1 of Chapter 2 of this book, or derived from the risk matrix approach. In the event of using risk graph methods, indicate the source and method of calibration of the method. Paragraph No ......

2.4 Check that the appropriate SIL table has been applied (high or low demand). Paragraph No ......

2.5 Review the target SIL(s) against the number of levels of protection and decide if a lower SIL target, with more levels of protection, is a more realistic design option. Paragraph No ......

2.6 Ensure that the design documentation, for example requirements specification, adequately identifies the use of the safety-related system for protection of the failure mode(s) defined. Paragraph No ......

3. Assessing the random hardware failure integrity of the proposed safety-related system

3.1 Create a reliability model(s), for example fault tree, block diagram, event tree, for the safety-related system and for the failure mode(s) defined. Paragraph No ......

3.2 Remember to address CCF in the above model(s). Refer to the literature for an appropriate model, for example BETA-PLUS. Paragraph No ......

3.3 Remember to quantify human error (where possible) in the above model(s). Paragraph No ......

3.4 Remember to address both auto and manual diagnostic intervals and coverage in the above model(s). Paragraph No ......

3.5 Select appropriate failure rate data for the model(s) and justify the use of sources. Paragraph No ......

3.6 Quantify the model(s) and identify the relative contributions to failure of the modules/components within the SRS (safety-related system). Paragraph No ......

4. Assessing the qualitative integrity of the proposed safety-related system

4.1 Check that the architectural constraints for the SIL in question have been considered and that the diagnostic coverage and safe failure fractions have been assessed. Paragraph No ......

4.2 Review each paragraph of Chapters 3 and 4 of this book HAVING REGARD TO EACH FAILURE MODE being addressed. Remember that the qualitative feature applies to the safety-related system for a SPECIFIC failure mode. Thus, a design review involving features pertaining only to 'spurious shutdown' would not be relevant where 'failure to shut down' is the issue. Paragraph No ......

4.3 Document which items can be reviewed within the organisation and which items require inputs from suppliers/subcontractors. Paragraph No ......

4.4 Obtain responses from suppliers/subcontractors and follow up as necessary to obtain adequate VISIBILITY. Paragraph No ......

4.5 Document the findings for each item above, and provide a full justification for items not satisfied but deemed to be admissible, for example non-use of Static Analysis at SIL 3 for a simple PLC. Paragraph No ......

4.6 Has the use of software downloaded from a remote location, and any associated problems, been addressed? Paragraph No ......

5. Reporting and recommendations

5.1 Prepare a draft assessment report containing, as a minimum:

- Executive summary
- Reason for assessment
- Definition of the safety-related system and its failure modes
- Calculation of target SIL
- Reliability model
- Assumptions inherent in reliability model, for example downtimes and proof-test intervals
- Failure data sources
- Reliability calculations
- Findings of the qualitative assessment

Report No ......

5.2 If possible include recommendations in the report as, for example:

> 'An additional mechanical relief device will lower the SIL target by one, thus making the existing proposal acceptable.'
>
> 'Separated, asynchronous PESs will reduce the CCF sufficiently to meet the target SIL.'

Paragraph No ......

5.3 Address the ALARP calculation where the assessed risk is greater than the broadly acceptable risk. Paragraph No ……

5.4 Review the draft report with the client and make amendments as a result of errors, changes to assumptions, proposed design changes etc.

Meeting (date) ……

6. Assessing vendors

6.1 In respect of the items, identified above, requiring the assessment to interrogate subcontractors/suppliers, take account of other assessments that may have been carried out, for example IEC 61508 assessment or assessment against one of the documents in Chapter 9 of this book. Review the credibility and rigour of such assessments. Paragraph No ……

6.2 In respect of the items, identified above, requiring the assessment to interrogate subcontractor/suppliers, ensure that each item is presented as formal evidence (document or test) and is not merely hearsay, for example 'a code review was carried out'. Paragraph No ……

7. Addressing capability and competence

7.1 Has a functional safety capability review been conducted as per Appendix 1. Paragraph No ……

7.2 Consider the competence requirements of Designers, Maintainers, Operators and Installers. Paragraph No ……

7.3 Establish the competence of those carrying out this assessment. Paragraph No ……

APPENDIX 3

BETAPLUS CCF MODEL, CHECKLISTS

1. CHECKLIST for Equipment containing Programmable Electronics

A scoring methodology converts this checklist into an estimate of BETA. This is available as the BETAPLUS software package.

(1) SEPARATION/SEGREGATION

Are all signal cables separated at all positions?

Are the programmable channels on separate printed circuit boards?

OR are the programmable channels in separate racks?

OR in separate rooms or buildings?

(2) DIVERSITY/REDUNDANCY

Do the channels employ diverse technologies?

1 electronic + 1 mechanical/pneumatic

OR 1 electronic or CPU + 1 relay based

OR 1 CPU + 1 electronic hardwired

OR do identical channels employ enhanced voting?
i.e. 'M out of N' where $N > M + 1$

OR $N = M + 1$

Were the diverse channels developed from separate requirements from separate people with no communication between them?

Were the 2 design specifications separately audited against known hazards by separate people and were separate test methods and maintenance applied by separate people?

(3) COMPLEXITY/DESIGN/APPLICATION/ MATURITY/EXPERIENCE

Does cross-connection between CPUs preclude the exchange of any information other than the diagnostics?

Is there >5 years' experience of the equipment in the particular environment?

Is the equipment simple <5 PCBs per channel?

OR <100 lines of code

OR <5 ladder logic rungs

OR <50 I/O and <5 safety functions?

Are I/O protected from overvoltage and overcurrent and rated >2:1?

(4) ASSESSMENT/ANALYSIS and FEEDBACK of DATA

Has a combination of detailed FMEA, fault tree analysis and design review established potential CCFs in the electronics?

Is there documentary evidence that field failures are fully analysed with feedback to design?

(5) PROCEDURES/HUMAN INTERFACE

Is there a written system of work on site to ensure that failures are investigated and checked in other channels? (Including degraded items which have not yet failed.)

Is maintenance of diverse/redundant channels staggered at such an interval as to ensure that any proof tests and cross-checks operate satisfactorily between the maintenance?

Do written maintenance procedures ensure that redundant separations as, for example, signal cables are separated from each other and from power cables and should not be re-routed?

Are modifications forbidden without full design analysis of CCF?

Is diverse equipment maintained by different staff?

(6) COMPETENCE/TRAINING/SAFETY CULTURE

Have designers been trained to understand CCF?

Have installers been trained to understand CCF?

Have maintainers been trained to understand CCF?

(7) ENVIRONMENTAL CONTROL

Is there limited personnel access?

Is there appropriate environmental control? (e.g. temperature, humidity)

(8) ENVIRONMENTAL TESTING

Has full EMC immunity or equivalent mechanical testing been conducted on prototypes and production units (using recognised standards)?

2. CHECKLIST AND SCORING for non-Programmable Equipment

Only the first three categories have different questions as follows:

(1) SEPARATION/SEGREGATION

Are the sensors or actuators physically separated and at least 1 metre apart?

If the sensor/actuator has some intermediate electronics or pneumatics, are the channels on separate PCBs and screened?

OR if the sensor/actuator has some intermediate electronics or pneumatics, are the channels indoors in separate racks or rooms?

(2) DIVERSITY/REDUNDANCY

Do the redundant units employ different technologies?
e.g. 1 electronic or programmable + 1 mechanical/pneumatic

OR 1 electronic, 1 relay based

OR 1 PE, 1 electronic hardwired

OR do the devices employ 'M out of N' voting where $N > M + 1$

OR $N = M + 1$

Were separate test methods and maintenance applied by separate people?

(3) COMPLEXITY/DESIGN/APPLICATION/ MATURITY/EXPERIENCE

Does cross-connection preclude the exchange of any information other than the diagnostics?

Is there >5 years' experience of the equipment in the particular environment?

Is the equipment simple, e.g. non-programmable type sensor or single actuator field device?

Are devices protected from overvoltage and overcurrent and rated >2:1 or mechanical equivalent?

(4) ASSESSMENT/ANALYSIS and FEEDBACK OF DATA

As for Programmable Electronics (see above)

(5) PROCEDURES/HUMAN INTERFACE

As for Programmable Electronics (see above)

(6) COMPETENCE/TRAINING/SAFETY CULTURE

As for Programmable Electronics (see above)

(7) ENVIRONMENTAL CONTROL

As for Programmable Electronics (see above)

(8) ENVIRONMENTAL TESTING

As for Programmable Electronics (see above)

The diagnostic interval is shown for each of the two (programmable and non-programmable) assessment lists. The **(C)** values have been chosen to cover the range 1–3 in order to construct a model which caters for the known range of BETA values.

For Programmable Electronics

Diagnostic coverage	Interval <1 min	Interval 1–5 mins	Interval 5–10 mins	Interval >10 mins
98%	3	2.5	2	1
90%	2.5	2	1.5	1
60%	2	1.5	1	1

For Sensors and Actuators

Diagnostic coverage	Interval <2 hrs	Interval 2 hrs–2 days	Interval 2 days–1 week	Interval >1 week
98%	3	2.5	2	1
90%	2.5	2	1.5	1
60%	2	1.5	1	1

THE BETAPLUS MODEL IS AVAILABLE, AS A SOFTWARE PACKAGE, FROM THE AUTHOR

APPENDIX 4

ASSESSING SAFE FAILURE FRACTION AND DIAGNOSTIC COVERAGE

In Chapter 3 Safe Failure Fraction was described and reference was made to two ways of assessing it.

1. By failure mode and effect analysis

Figure A4.1 shows an extract from a failure mode and effect analysis (FMEA) covering a single failure mode (e.g. OUTPUT FAILS LOW).

> Columns (A) and (B) identify each component.
> Column (C) is the total part failure rate of the component.
> Column (D) gives the failure mode of the component leading to the failure mode (e.g. FAIL LOW condition).
> Column (E) = Column (D) × (E) shows the appropriate proportion of column (C) (e.g. 20% for U8).
> Column (F) shows the assessed probability of that failure being diagnosed. This would ideally be 100% or 0 but a compromise is sometimes made when the outcome is not totally certain.
> Column (H) is a working column which multiplies the mode failure rate by the diagnostic coverage for each component.

Cells at the bottom of the spreadsheet in Figure A4.1 contain the algorithms to calculate diagnostic coverage (63%) and SFF (92%).

XYZ MODULE							
A	B	C	D	E	F	G	H
COMP REF	**DESCRIPTION**	**F.Rate pmh**	**MODE 1**	**M1 F.Rate pmh**	**% Diag M1**	**NOTES**	**E * F**
U6/7	2 × MOS LATCHES @ 0.01	0.02	20%	0.004	90		0.0036
U8	PROG LOGIC ARRAY	0.05	20%	0.01	0		0
U9-28	20 × SRAM @ .02	0.4	20%	0.08	90		0.072
U29-31	4 × FLASH MOS 4M @ .08	0.32	20%	0.064	50		0.032
TR23	npn lp	0.04	S/C	0.012	0		0
TOTAL F.Rate		0.83		0.17			0.1076
WEIGHTED %					**63**		
SFF =					**92**		

Figure A4.1
FMEA

Diagnostic coverage is obtained from the sum of column H divided by the sum of column E.

SFF is obtained by taking the proportion of diagnosed hazardous failures (total of Column E × the diagnostic coverage) PLUS the failures deemed to be non-hazardous in this context (total of Column C minus total of Column E). This total is divided by the total failure rate (Column C) to obtain the SFF.

Typically this type of analysis requires 4 mandays of effort based on a day's meeting for a circuit engineer, a software engineer who understands the diagnostics and the safety assessor carrying out the 'component by component' review. A further day allows the safety assessor to add failure rates and prepare the calculations and a report.

2. By a block level assessment of the architecture

The following example is much less rigorous than the above FMEA and requires say three persons (as above) for approximately 3 hours. It is, however, only suitable for demonstrating 60% SFF. In this example over 90% diagnostic coverage is assessed.

Assume that the hazardous failures are assessed to be 80% of the total then SFF is:

$$[91\% \times 80\%] + 20\% = 92.8\%$$

and thus an SFF in excess of 60% can be claimed with reasonable confidence.

Consider a non-redundant configuration where 60% fault detection coverage is required for SIL 1 (Table B). A design review was carried out at XYZ Ltd. The circuitry, together with the watchdog and software diagnostic arrangements, was examined and the following was established:

(a) A single hardware watchdog is provided, by means of a diode pump and discharge circuit. This establishes a minimum (not a maximum) reset frequency. A number (approximately six) of other software watchdogs monitor specific tasks (e.g. bus, communications, database, function block update) and need to be satisfied or else they trip the hardware watchdog. Thus the buses have an extremely high diagnostic coverage.

(b) Each function block generates a checksum which is computed each time the block is written.

(c) Built-in software diagnostics ('executive crash') interrogates a number of logical scenarios within the processing and also cause watchdog trip if not satisfied.

(d) The watchdog trip will result in the I/P low and slamshut relays released modes. (Note: this should be confirmed with the system electrical design.) Watchdog trip may be followed by one of three 'hard switched' options:

- Manual restart only
- Warm restart using previous parameter setting
- Cold restart involving download of program from EPROM

(Note: check that the manual restart option is selected.)

(e) The checksum (b) diagnoses all single bit failures. Multiple bit errors are usually spread in memory in such a way that they will be exposed to more than one of these block checksums thus increasing the probability of detection. If this does not occur then there remains the possibility that an error will lead to an illegal jump which may well invoke the diagnostic software (c). A better than 95% diagnostic coverage is thus assessed for MEMORY.

(f) CPUs in both the CPU board and I/O boards will trip the watchdog in respect of slow running. High running, whilst not directly detected by the watchdog, will likely violate the memory timings and serial bus timings thus tripping a software watchdog (b) or diagnostics (c). Better than 90% diagnostic coverage is thus assessed for CPUs.

(g) Approximately 99% diagnostic is assessed for the PSUs since incorrect levels are directly detected by the watchdog (a).

(h) Input/Output buses are directly interrogated by a software watchdog (c). In the case of input channels hardware failures at the input are directly signalled to the software watchdog (c) giving better than 80% diagnostics. In the case of output channels the output channel can be fed back to an input as a diagnostic feature giving better than 80% diagnostics.

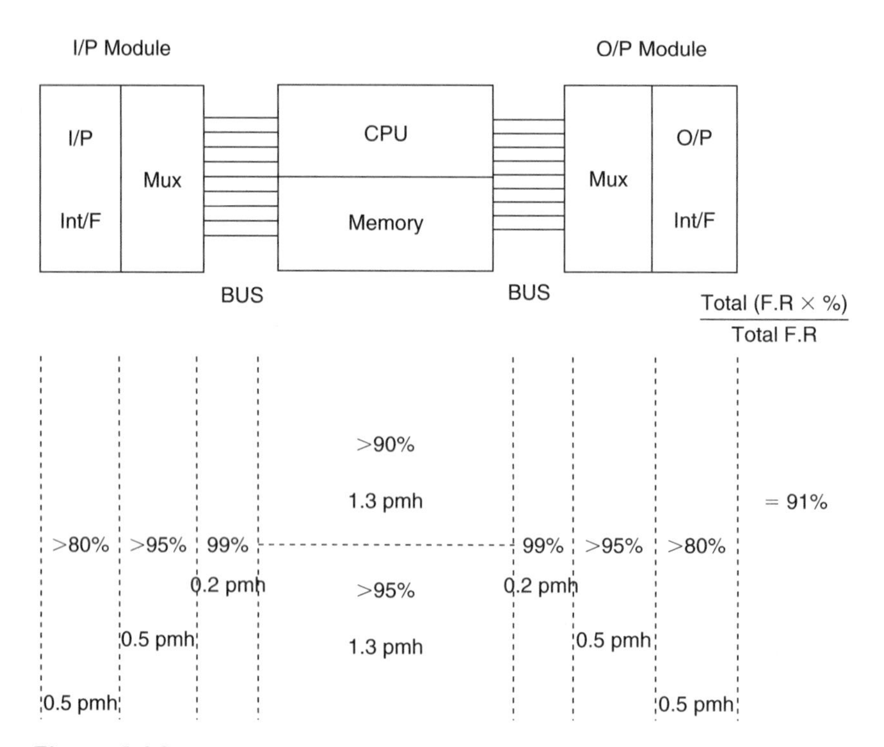

Figure A4.2
Block level assessment

(i) MUXs are believed to be between 90% and 99% covered by diagnosis.

Figure A4.2 shows how the foregoing has been used to assess diagnostic coverage (%) for each block of the architecture. The percentages have been weighted by failure rate and summed to provide an overall assessment.

APPENDIX 5

ANSWERS TO EXAMPLES

Answer to Exercise 1 (Chapter 2.2)

Maximum tolerable failure rate leading to fatality is 10^{-5}pa/10^{-1} = 10^{-4}pa
However, the actual process failure rate is 0.05 pa = 5×10^{-2}pa
Thus the protection system should have a target probability of failure on demand (PFD) no worse than:

$$10^{-4}\text{pa}/5 \times 10^{-2}\text{pa} = 2 \times 10^{-3}$$

The target is dimensionless and is thus a PFD. The low demand column in Table 1.1 is therefore indicated
Thus the requirement is SIL 2

Answer to Exercise 2 (Chapter 2.2)

Answer 2.1

Since there are 10 sources of risk (at the same place) the maximum tolerable fatality rate (per risk) is $10^{-5}/10 = 10^{-6}$pa

Target toxic spill rate is 10^{-6}pa/ 10^{-1} = 10^{-5}pa

However, the actual spill rate is 1/50 pa = 2×10^{-2}pa
Thus the protection system should have a target probabilty of failure on demand no worse than:

$$10^{-5}\text{pa}/2 \times 10^{-2}\text{pa} = 5 \times 10^{-4}$$

The target is dimensionless and is thus a PFD. The low demand column in Table 1.1 is therefore indicated
Thus the requirement is SIL 3

Answer 2.2

The additional protection reduces the propagation to fatality to 1:30 so the calculation becomes:

Target spill rate is 10^{-6}pa/3.3×10^{-2}pa $= 3 \times 10^{-5}$pa

However, spill rate is 1/50 pa $= 2 \times 10^{-2}$pa

Thus the protection system should have a target probabilty of failure on demand no worse than:

3×10^{-5}pa/2×10^{-2}pa $= 1.5 \times 10^{-3}$

Thus the requirement is SIL 2 (low demand)

Answer to Exercise 3 (Chapter 2.2)

Target maximum tolerable risk $= 10^{-5}$pa
Propagation of incident to fatality $= 1/200 = 5 \times 10^{-3}$
Thus target maximum tolerable failure rate $= 10^{-5}$pa/$5 \times 10^{-3} = 2 \times 10^{-3}$pa
Note 2×10^{-3}pa $= 2.3 \times 10^{-7}$ per hour
The requirement is expressed as a rate, thus the high demand column of Table 1.1 is indicated at SIL 2

Answer to Exercise 4 (Chapter 2.2)

Repeat Exercise 1 using:

(a) The risk graph (Figure 2.3)
1 death (say V = 0.5), thus C = 0.5; Exposure >0.1 (frequent); Avoidance <10% (not likely); Demand rate 0.05 pa (middle column), suggests:
SIL 3
Note that the result is one SIL more onerous than the previous approach.

Answer to Exercise 5 (Chapter 2.3)

For the expense to just meet the cost per life saved criterion then:

£2 000 000 = £proposal/$(8 \times 10^{-6} - 2 \times 10^{-6}) \times 3 \times 25$ = £900

Thus an expenditure of £900 would be justified if the risk reduction can be obtained for this outlay. Expenditure greatly in excess of this could be argued to be grossly disproportionate to the benefits.

Answer to Exercise (Chapter 11)

Paragraph 11.2 Protection system

The target Unavailability for this 'add-on' safety system is therefore 10^{-5}pa/2.5×10^{-3}pa = $\mathbf{4 \times 10^{-3}}$ which indicates **SIL 2**

Paragraph 11.4 Reliability block diagram

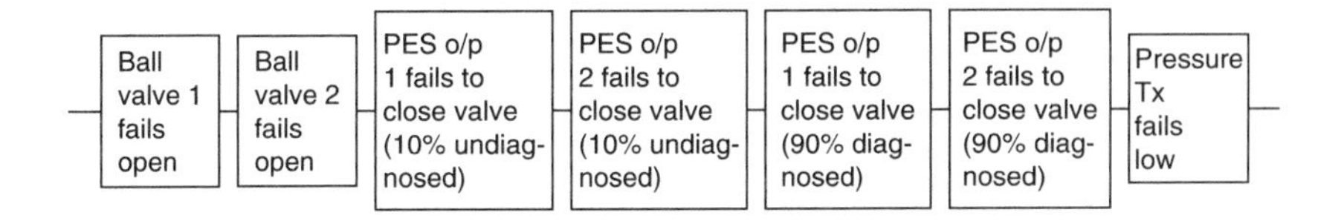

Paragraph 11.6 Quantifying the model

(a) Ball valve SS1 fails open.
Unavailability = λ MDT = $0.8 \times 10^{-6} \times 4000 = 3.2 \times 10^{-3}$

(b) Ball valve SS2 fails open.
Unavailability = λ MDT = $0.8 \times 10^{-6} \times 4000 = 3.2 \times 10^{-3}$

(c) PES output 1 fails to close valve (undiagnosed failure).
Unavailability = 10% λ MDT = $0.025 \times 10^{-6} \times 4000$
$= 1 \times 10^{-4}$

(d) PES output 2 fails to close valve (undiagnosed failure).
Unavailability = 10% λ MDT = $0.025 \times 10^{-6} \times 4000$
$= 1 \times 10^{-4}$

(e) PES output 1 fails to close valve (diagnosed failure).
Unavailability = 90% λ MDT = $0.225 \times 10^{-6} \times 4 = 9 \times 10^{-7}$

(f) PES output 2 fails to close valve (diagnosed failure).
Unavailability = 90% λ MDT = $0.225 \times 10^{-6} \times 4 = 9 \times 10^{-7}$

(g) Pressure transmitter fails low.
Unavailability = λ MDT = $0.5 \times 10^{-6} \times 4000 = 2 \times 10^{-3}$

The predicted Unavailability is obtained from the sum of the unavailabilities in (a) to (g) = 8.6×10^{-3}.
(Note: the estimate of $1.2 \times 10^{-2} \times$ in Chapter 8.)
This is higher than the unavailability target. The argument as to the fact that this is still within the SIL 2 target was discussed in Chapter 3.3.3. We chose to calculate an unavailability target and thus it is NOT met.

74% from items (a) and (b) the valves.
23% from item (g) the pressure transmitter.
Negligible from items (c)–(f) the PES.

Paragraph 11.7 Revised diagrams

Reliability block diagram

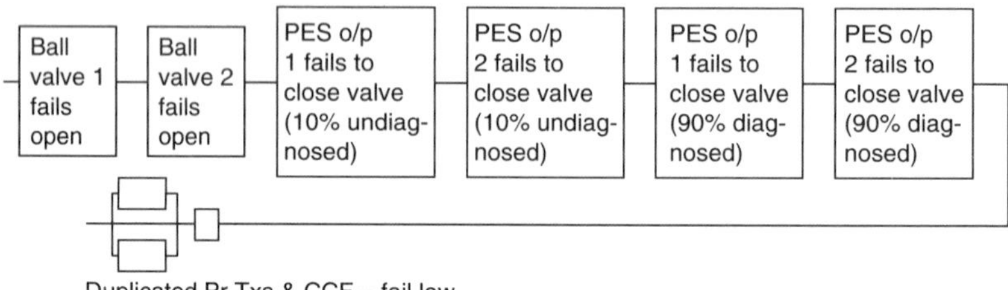

Equivalent fault tree

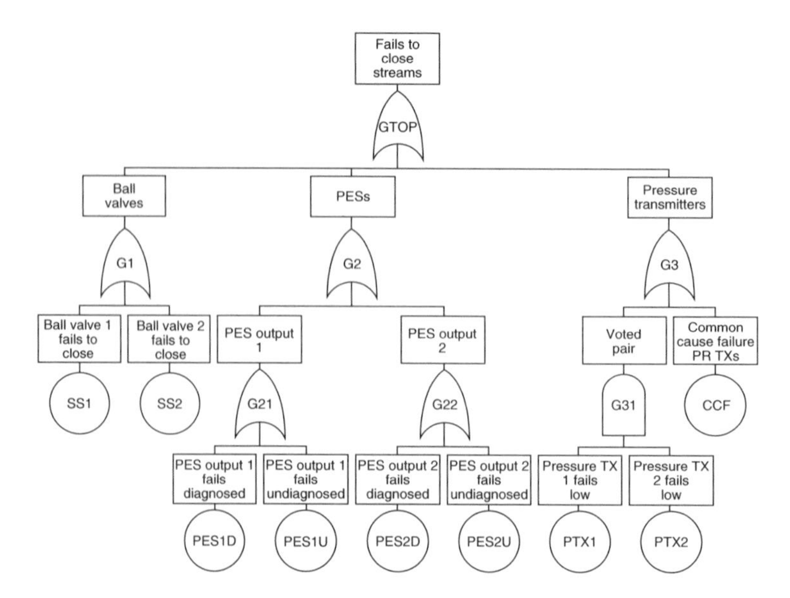

Paragraph 11.9 Quantifying the revised model

Changed figures are shown in bold.

(a) Ball valve SS1 fails open.
Unavailability = λ MDT = $0.8 \times 10^{-6} \times \mathbf{2000} = \mathbf{1.6 \times 10^{-3}}$

(b) Ball valve SS2 fails open.
Unavailability = λ MDT = $0.8 \times 10^{-6} \times \mathbf{2000} = \mathbf{1.6 \times 10^{-3}}$

(c) PES output 1 fails to close valve (undiagnosed failure).
Unavailability = 10% λ MDT = $0.025 \times 10^{-6} \times \mathbf{2000}$
$= \mathbf{5 \times 10^{-5}}$

(d) PES output 2 fails to close valve (undiagnosed failure).
Unavailability = 10% λ MDT = $0.025 \times 10^{-6} \times \mathbf{2000}$
$= \mathbf{5 \times 10^{-5}}$

(e) PES output 1 fails to close valve (diagnosed failure).
Unavailability = 90% λ MDT = $0.225 \times 10^{-6} \times 4 = 9 \times 10^{-7}$

(f) PES output 2 fails to close valve (diagnosed failure).
Unavailability = 90% λ MDT = $0.225 \times 10^{-6} \times 4 = 9 \times 10^{-7}$

(g) Voted pair of pressure transmitters.
Unavailability = $\boldsymbol{\lambda^2 T^2/3 = [0.5 \times 10^{-6}]^2 \times 4000^2/3}$
$= \mathbf{1.3 \times 10^{-6}}$

(h) Common cause failure of pressure transmitters.
Unavailability = **9%** $\boldsymbol{\lambda}$ **MDT** = $\mathbf{0.09 \times 0.05 \times 10^{-6} \times 2000}$
$= \mathbf{9 \times 10^{-5}}$

The predicted Unavailability is obtained from the sum of the unavailabilities in (a) to (h) = $\mathbf{3.3 \times 10^{-3}}$ which meets the traget.
(Note: the estimate of 5.1×10^{-3} in Chapter 8.)

Paragraph 11.10 ALARP

Assume that further improvements, involving CCF and a further reduction in proof-test interval, can be achieved for a total cost of £1000. Assume, also, that this results in an improvement in unavailability, of the safety-related system, from $\mathbf{3.3 \times 10^{-3}}$ to the PFD associated with the Broadly Acceptable limit of $\mathbf{4 \times 10^{-4}}$. It is necessary to consider, applying the ALARP principle, whether this improvement should be implemented.

If the target unavailability of $\mathbf{4 \times 10^{-3}}$ represents a maximum tolerable risk of $\mathbf{10^{-5}}$**pa** then it follows that $\mathbf{3.3 \times 10^{-3}}$ represents a risk of $10^{-5} \times 3.3/4 = \mathbf{8.3 \times 10^{-6}}$**pa**. If $\mathbf{10^{-6}}$**pa** is taken as the boundary of the negligible risk then the proposal remains within the tolerable range and thus subject to ALARP.

Assuming a two fatality scenario, the cost per life saved over a 40-year life of the equipment (without cost discounting) is calculated as follows:

$\mathbf{3.3 \times 10^{-3}}$ represents a risk of 8.3×10^{-6}
$\mathbf{4 \times 10^{-4}}$ represents a risk of 10^{-6}

Cost per life saved = £1000/(40 × 2 lives × [8.3 − 1] 10^{-6}) = **£1 700 000**

On this basis, if the cost per life saved criterion were £1 000 000, then justification for the further improvement would be considered marginal as the benefit is just below (but close to) the criteria. On the other hand it would be justified if the criterion were £2 000 000.

Paragraph 11.11 Architectural constraints

(a) PES
The safe failure fraction for the PESs is given by 90% diagnosis of 5% of the failures, which cause the failure mode in question, PLUS the 95% which are 'fail safe'
Thus (90% × 5%) + 95% = 99.5%
Consulting the tables in Chapter 3.3.2 then:
If the simplex PES is regarded as Type B then SIL 2 can be considered if this design has >90% safe failure fraction

(b) Pressure transmitters
The safe failure fraction for the transmitters is given by the 75% which are 'fail safe'
If they are regarded as Type A then SIL 2 can be considered since they are voted and require less than 60% safe failure fraction
Incidentally, in the original proposal, the simplex pressure transmitter would not have met the architectural constraints

(c) Ball valves
The safe failure fraction for the valves is given by the 90% which are 'fail safe'
If they are regarded as Type A then SIL 2 can be considered since they require more than 60% safe failure fraction

Comments on Example (Chapter 12)

The following are a few of the criticisms which could be made of the Chapter 12 report.

12.2 Integrity requirements

In Chapter 11 the number of separate risks to an individual was taken into account. As a result the 10^{-4}pa target was amended to 10^{-5}pa. This may or may not be the case here but the point should be addressed.

12.4.1 ALARP

It was stated that nothing could be achieved for £672. It may well be possible to achieve significant improvement by reducing proof-test intervals for a modest expenditure.

12.5 Failure rate data

It is not clear how the common cause failure proportion has been chosen. This should be addressed as in Chapter 11.

Other items

(a) There is no mention of the relationship of the person who carried out the assessment to the provider. Independence of the assessment needs to be explained.
(b) Safe failure fraction was not addressed.
(c) Although the life-cycle activities were referred to, the underlying function safety capability of the system provider was not called for.

Appendix 6

References

Carey M, Proposed framework for addressing human factors in IEC 61508, Amey VECTRA Ltd.

DIN V 19 250, Measurement and control, fundamental safety aspects for measuring and control protective equipment.

DIN VDE 0801, 1990, Principles for computers in safety-related systems.

EEMUA Guidelines – Publication No 160, 1989, Safety related instrument systems for the process industry (including programmable electronic systems).

EN 50126 Draft European Standard: Railway applications – The Specification and Demonstration of Dependability, Reliability, Maintainability and Safety (RAMS).

EN 50128 – Software for railway control and protection systems.

EN 50129 – Hardware for railway control and protection systems.

EN 60204-1 Safety of machinery – electrical equipment of machines.

EN 9541-1 Safety of machinery in safety-related parts of control systems.

Gulland W G, Repairable redundant systems and the Markov fallacy, *Journal of Safety and Reliability Society* Vol 22 No 2 Summer 2002.

HSE, 1992, Tolerability of risk for nuclear power stations, UK Health and Safety Executive, ISBN 0 1188 6368 1. *Often referred to as TOR.*

HSE, 2001, Reducing risks, protecting people. *Often referred to as R2P2.*

HSE, 1995, Out of control: control systems: why things went wrong and how they could have been prevented, HSE Books ISBN 0 7176 0847 6.

HSE 190, 1999, Preparing safety reports: control of Major Accident Regulations. Appendix 4 addresses ALARP.

HSE, 2000, Regulating higher hazards: exploring the issues.

HSE Publication, 1989, Guidance on the use of Programmable Electronic Systems in safety-related applications.

IEC Standard 61508, 2000, Functional safety: safety related systems – 7 parts.

IEC Standard 61713, 2000, Software dependability through the software life-cycle processes – application guide.

IEC Draft Standard 62061, Safety of machinery – functional safety of electronic and programmable electronic control systems for machinery.

IEC Draft International Standard 61511 (2003): Functional safety – safety instrumented systems for the process industry sector.

IEC Draft International Standard 61513: Nuclear Power Plants – Instrumentation and control for systems important to safety – general requirements for systems.

IEC Publication 61131, Programmable controllers, 8 Parts (Part 3 is programming languages).

IEE Publication SEMSPLC, 1996, Safety-related application software for Programmable Logic Controllers, ISBN 0 8529 6887 6.

IEE Publication, 1992, Guidelines for the documentation of software in industrial computer systems, 2nd edition, ISBN 0 8634 104 664.

IEE/BCS, 1999, Competency guidelines for safety-related system practitioners, ISBN 0 8529 6787 X.

Institution of Gas Engineers & Managers publication IGE/SR/15, Programmable equipment in safety related applications, Edition 3 (1998) and Amendments (2000 & 2002) ISSN 0 367 7850.

Instrument Society of America, S84.01, 1996, Application of safety instrumented systems for the process industries, ISBN 1 5561 7590 6.

MISRA (Motor Industry Research Assoc), 1994, Development guidelines for vehicle based software.

Norwegian Oil Ind Assoc, OLF-077, Recommended guidelines for the application of IEC 61508 in the petroleum activities on the Norwegian Continental Shelf.

RTCA DO-178B/(EUROCAE ED-12B), 1992, Software considerations in airborne systems and equipment certification.

Simpson K G L, Reliability assessments of repairable systems – is Markov modelling correct? *Journal of Safety and Reliability Society*, Vol 22 No 2 Summer 2002.

Smith D J, 2000, *Reliability, Maintainability and Risk*, 6th Edition (Butterworth Heinemann UK) ISBN 0 7506 5168 7.

Smith D J, *FARADIP.THREE, Version 4.1, 1999, User's manual*, Reliability software package ISBN 0 9516562 3 6.

Smith D J, *BETAPLUS Version 1.0, 1997, User's manual*, Common cause failure software package ISBN 09516562 5 2.

Smith D J, 2000, *Developments in the Use of Failure Rate Data and Reliability Prediction Methods for Hardware* ISBN 09516562 6 0.

Storey N, 1996, *Safety Critical Computer Systems*, Addison Wesley, ISBN 0 2014 2787 7.

Technis Guidelines Q124, 2004, Demonstration of Product/System Compliance with IEC 61508.

UKAEA, 1995, Human reliability assessors guide (SRDA-R11), June 1995, Thomson House, Risley, Cheshire WA3 6AT ISBN 0 8535 6420 5.

UK MOD Interim Defence Standard 00-55: The procurement of safety critical software in defence equipment.

UK MOD Interim Defence Standard 00-56: Hazard analysis and safety classification of the computer and programmable electronic system elements of defence equipment.

UK MOD Interim Standard 00-58: A guideline for HAZOP studies on systems which include programmable electronic systems.

UK MOD Interim Defence Standard 00-54: Requirements for safety-related electronic hardware in defence equipment.

UKOOA: Guidelines for Process Control and Safety Systems on Offshore Installations.

UL (Underwriters Laboratories Inc, USA), 1998, Software in programmable components, ISBN 0 7629 0321 X.

VDE 0801 – Principles for computers in safety-related systems.

Wichmann B, Validation of measurement software, National Physical Laboratory, Draft 2002.

APPENDIX 7

'HIGH AND LOW DEMAND'

1. Showing the equivalence of the low and high demand tables

EUC	SRS

Maximum Tolerable Risk = MTR (fatalities per annum)
Propagation to fatality = P
Maximum Tolerable Failure rate of the TOTAL system (EUC and SRS combined) = MTR/P (failures per annum) = $\lambda_{\mathbf{tot}}$

The demand rate on the SRS is the failure rate of the EUC = $\lambda_{\mathbf{dem}}$

The max tolerable PFD target for the SRS is thus $\lambda_{\mathbf{tot}}/\lambda_{\mathbf{dem}}$

At this point you would normally consult the low demand PFD table for the SIL

BUT:

This PFD (namely $\lambda_{tot}/\lambda_{dem}$) = $\lambda_{srs} \times$ PTI/2 where λ_{srs} is the failure rate we are looking for for the SRS for high demand table (where PTI is the proof-test interval).

The mean time to the next demand is in fact the reciprocal of the demand rate (because they are random demands).

SO:

PFD = $\lambda_{srs} \times$PTI/2 = $\boldsymbol{\lambda_{srs} \times 1/\lambda_{dem}}$

BUT we have shown above that the PFD is $\lambda_{tot}/\lambda_{dem}$

Therefore $\lambda_{srs} = \lambda_{tot}$ for the purposes of the high demand table. This is what we expect since the Max Tolerable Failure rate will be that of the SRS. This is because we are effectively assuming the EUC is 'always' failed.

2. Applying common sense

2.1 If the demand is small then we use the PFD (low demand table) as achieved by any realistic combination of λ_{srs} and PTI.

2.2 If the demand is very high then there is no question of a proof-test interval because the demands are continuous, and so we use λ_{srs} (high demand table).

2.3 Since PTIs of less than 3 months are unlikely to be realistic then a borderline demand rate of something like 4 pa, to differentiate between the two cases, would seem to be sensible.

APPENDIX 8

SOME TERMS AND JARGON OF IEC 61508

The seven 'Parts' of IEC 61508 are described as '**normative**' which means they are the Standard proper and contain the requirements which should be met. Some of the annexes, however, are described as '**informative**' in that they are not requirements but guidance which can be used when implementing the normative parts. It should be noted that the majority of Parts 5, 6 and 7 of the Standard are informative annexes.

A few other terms are worth a specific word or so here:

Functional safety is the title of this book and of IEC 61508. It is used to refer to the reliability (known as integrity in the safety world) of safety-related equipment. In other words it refers to the probability of it functioning correctly, hence the word 'functional'.

E/E/PE (Electrical/Electronic/Programmable Electronic Systems) refers to any system containing one or more of those elements. This is taken to include any input sensors, actuators, power supplies and communications highways. Providing that one part of the safety-related system contains one or more of these elements the Standard is said to apply to the whole.

EUC (Equipment under control) refers to the items of equipment which the safety-related system being studied actually controls. It may well be, however, that the EUC is itself safety related and this will depend upon the SIL calculations described in Chapter 3.

HR and R are used (in IEC 61508) to refer to 'Highly Recommended' and 'Recommended'. This is a long winded way of saying that HR implies activities or techniques which are deemed necessary at a particular SIL and for which a reasoned case would be needed for not employing them. R implies activities or techniques which are deemed to be 'good practice'.

NR is used to mean 'Not Recommended', meaning that the technique is not considered appropriate at that SIL.

Verification and Validation: Verification (as opposed to Validation) refers to the process of checking that each step in the life-cycle meets earlier requirements. Validation (as opposed to Verification) refers to the process of checking that the final system meets the original requirements.

Type A components (hardware or software) implies that they are well understood in terms of their failure modes and that field failure data is available. See Section 3.3.2.
Type B components (hardware or software) implies that any one of the Type A conditions is not met. See Section 3.3.2.

Should/Shall/Must: In standards work the term 'must' usually implies a legal requirement and has not been used in this book. The term 'shall' usually implies strict compliance and the term 'should' implies a recommendation. We have not attempted to differentiate between those alternatives and have used 'should' throughout this book.

Index